Springboards for Budding Preachers

Open-Air Preaching Transcripts

Ray Comfort
With Emeal Zwayne

*Our sincere thanks to Judy Notchick,
Jaylene Daugherty, and Anita Alvarado
for typing these transcripts, and to
Melissa Day and the staff of Living Waters
for their editorial help*

Introduction

To understand the principles conveyed in these messages, we strongly suggest that you be familiar with two foundational teachings: "Hell's Best Kept Secret" and "True and False Conversion." You can freely listen to these online at www.livingwaters.com.

To learn more about open-air preaching—such as how to attract a crowd, handle hecklers, etc.—see *The Evidence Bible* (Bridge-Logos Publishers). For actual open-air preaching footage, see "Open-Air Preaching: Four-in-One." This DVD contains four programs:

- How to Preach Open-Air
- In Season, Out of Season
- Open-Air Preaching: New York
- Open-Air Preaching from A to Z

In this DVD, join Ray Comfort as he takes a team from David Wilkerson's church and open-air preaches in Washington Square, New York, right in the middle of a Hare Krishna convention. Witness an encounter with the New York police, learn how to draw a crowd using a fake funeral, and discover how to handle "hecklers." Then go step-by-step through open-air preaching at UCLA and other Southern California universities. Also join Ray and his team as they preach open-air in Paris, Jerusalem, London, Tokyo, Santa Monica, New Zealand, and Amsterdam. These resources are available at www.livingwaters.com.

Contents

Introduction

There is one God who's perfect, holy, and righteous. You have violated God's Law. You cannot guarantee another breath. Have you loved God with heart, mind, soul, and strength?" Down through the ages, God raised up such men at George Whitefield, John Wesley, and others to preach where people gathered. "*Come to your senses!*" In the open-air. "*God is rich in mercy to all that call upon Him!*"

Charles Spurgeon said, "No sort of defense is needed for preaching out of doors. But it would need very potent arguments to prove that a man had done his duty who has never preached beyond the walls of his meeting place." "*Think soberly about your eternal salvation. Repent and put your trust in Jesus Christ. You'll pass from death into life, darkness into light. Instead of God being afar off, He'll be your closest friend. He'll seal you to His Holy Spirit and make you a new person. You'll be born again with a new heart and new desires.*"

It is our conviction that God is raising up men and women who will faithfully preach truths that are calculated to bring about revival. Men and women who are not afraid to lift up their voice like a trumpet and show this people their transgressions. "*God became a human being in Jesus Christ and this perfect, sinless Man gave His life as a sacrifice for the sin of the world!*"

Hi, this is Ray Comfort. For twelve years I preached almost daily in "Speaker's Corner" in the city of Christchurch, New Zealand. During those years, I learned that there were certain principles that should be used to be effective when open-air preaching: How to draw a crowd, how to begin speaking, how to hold their attention, what subjects to cover. But a number of years ago, my son-in-law, EZ, and I began preaching outside the local courthouse here in the city of Bellflower, California. The dynamic changed completely from what I was used to. We didn't have to draw a crowd—they were already there. Each day between forty and sixty people lined up waiting to see the judge and they couldn't get away.

The area in which we spoke was public domain and the local sheriff personally assured me that I had a First Amendment right to be there. So EZ and I preached at the courts almost every day for two-and-a-half years. One day I asked our soundman, Scotty, to begin recording the preaching. So he showed up wearing these large earphones. He had a boom pole stretched up about six feet with a shotgun microphone on the top. He's wearing a battery pack, wires everywhere, and he sat right in the middle of the crowd in the line. Someone remarked, "Now there goes an extremely deaf person." It was really funny. And as time went on he became more discreet as he recorded us.

Ironically, after thirty days of recording, two police officers stopped me from speaking and issued me a court order. They said that a judge had declared that it was no longer legal to speak there. At the end of the last message, you'll hear them pull me aside and issue me a court order. It was an obvious infringement on our First Amendment rights. We knew that the judge had no right to do that. So at present our lawyers are suing him for twenty-five dollars, and said that we'll be back there preaching very soon.

In the meanwhile, Scotty and I went to our local DMV and found even more people lined up, and we began preaching there. The California Highway Patrol told us that we needed a permit to hand out tracts, but they amazingly said that it was our First Amendment right to preach there. So being stopped at the courts actually caused us to find another place to preach. You might like to check at your own courts or your own local DMV as potential places for you to preach.

It's our hope that God will use these recordings to encourage and equip you. Each message begins with a very polite introduction, who we are, etc. We've left one or two introductions in for you to hear. Then the message begins in the natural realm, using a thought-provoking springboard. This is an "attention-getter." Then we spring to the spiritual, go through the Law, opening up the Commandments, into Judgment Day, the reality of Hell, into the cross, the resurrection, repentance, the necessity of faith, then a call to immediate action. Always remember you're heading for the cross.

To conclude, we always thank them for their attention and offer a free "What Hollywood Believes" CD. You can hear this message on www.whathollywoodbelieves.com/listen.shtml. By the way, you'll notice footsteps walking past Scotty each day at a certain point during the preaching. This is the court staff arriving for work.

In the following transcripts (also available at www.livingwaters.com/springboards), you'll find biblical references for Scriptures used in each message. Please feel free to use these springboards and make them your own; they're not copyrighted.

It's vital to enrich the gospel with the Word of God because Scripture has the promise that it won't return void. It's quick and powerful, sharper than any two-edged sword, and cuts through to the marrow of the bones.

Thank you so much for your concern for the lost. May God bless you as you listen.

You've Violated God's Law

Some of you are here today because you have allegedly violated man's law. The Bible alleges that both you and I have violated *God's* Law; that is, the Ten Commandments. So what I'm going to do is put you on the stand, just for a few moments, and examine you under the light of that Law. You can make the determination of your innocence or your guilt, and what your eternal destiny will be, whether it be a place called Heaven or a place called Hell. So this is a very important issue and I'd be grateful for your attention.

Let's begin with the Ninth Commandment, "You shall not bear false witness."[1] Have you ever told a lie? Are you innocent or guilty of breaking that Commandment? Or the Eighth, "You shall not steal."[2] Have you ever stolen something, irrespective of its value? Or the Third Commandment, "You shall not take the name of the Lord your God in vain."[3] Have you ever used God's name as a cussword to express disgust—something called blasphemy? Or the Seventh Commandment, "You shall not commit adultery."[4] But, guys, listen to what Jesus said of that Commandment. He said, "But I say to you, whoever looks upon a woman to lust after her has committed adultery already with her in his heart."[5] Have you ever done that? You say, *Man, have I ever done that? That's all I ever do. I'm guilty of that one.* In fact, you say, "You know I'm guilty of breaking those *four* Commandments." Well, if that's true, on the Day of Judgment, God will see you as a lying, thieving, blasphemous, adulterer at heart—very serious crimes in His sight. And the Bible makes it very clear your eternal destiny will not be Heaven, but a place called Hell. The Scriptures warn, "All liars will have their part in the lake of fire."[6] No thief, no adulterer, no fornicator (those who've had sex out of marriage) will inherit the Kingdom of God.[7]

You say, "You don't have to worry about me. I'll be fine. I won't end up in Hell, because I confessed my sins to God. I tell Him I'm sorry." Folks, that won't help you on the Day of Judgment. That's like saying to the judge today, "Hey, Judge, I confess I committed the crime. Sorry." He'd probably say, "So you're confessing your guilt. You *should* be sorry. You've done wrong."

No, you need someone who can pay your fine. And, folks, that's what God did for you and I two thousand years ago. The Bible says He sent forth His Son, made of a woman,[8] born of a virgin.[9] Do you know what that means? That means that Jesus of Nazareth was different than you and I. He was morally perfect. He was without sin, and this perfect Man gave His perfect life as a perfect sacrifice for the sin of the world.

---

1 Exodus 20:16
2 Exodus 20:15
3 Exodus 20:7
4 Exodus 20:14
5 Matthew 5:27,28
6 Revelation 21:8
7 See 1 Corinthians 6:9.
8 See Galatians 4:4.
9 Matthew 1:23

Now you know that, because you live in America. You know that Christ died for your sins. We just celebrated Christmas a couple of months ago—the birth of the Savior. We're celebrating Easter coming up soon—the death and resurrection of the Son of God. We know that Christ died for our sins, but what does it mean to you as an individual? What are the implications?

Let me put it this way. Do you realize that you're on death row—that you're waiting to die; that you're in a holding cell? Normally it has a nice blue roof, good lighting, good air-conditioning—but this life is a holding cell. Ten out of ten die, and the reason you and I will die is because we've violated God's Law. God said, "The soul that sins, it shall die."[10] The wages of sin is death,[11] and the proof that you've sinned against God will be your death.

But the same God that the Bible says is the Judge of the Universe is rich in mercy and He sent His Son to suffer and die on the cross on our behalf—to take the punishment for our sins; to pay the fine in His life's blood. That means God can commute your death sentence. He can nullify your death sentence. He can allow you to live, because of what Jesus did on the cross. The Bible says, "For God so loved the world that He gave His only begotten Son that whoever believes in Him should not perish but have everlasting life."[12] Through the death and resurrection of Jesus Christ, God can now dismiss your case. He can forgive your sins. He can allow you to live. So what must you do?

The Bible says that there are two things you must do. One is, you *must* repent—turn from your sins. Don't just confess your sins to God—confess and *forsake* them; and don't just "believe" in Jesus—*trust* in Him, like you trust a parachute to save you. You don't "believe" in a parachute. You put your faith *into* it. You put your *trust* in it. And that's what you must do with the Savior. And, folks, please do it today. Don't wait until tomorrow. You don't know if you'll be here tomorrow. The only thing you can guarantee for sure is the air going into your lungs at this present moment. You cannot guarantee another breath. That comes by the grace and mercy and patience of God.

Every day 150,000 people die.[13] They're swallowed by death. People, like you and me. People with the same desires, same concerns and the same plans for the future. We'll be snatched into eternity one day, and today could be your turn. Your number could be up. Mine could be up. So please consider seriously your eternal salvation.

⋆ ⋆ ⋆

California Girl

The other day I spoke to a young Korean girl. She was dressed kind of different and I wrongly surmised that she would speak to me in broken English. She didn't. She spoke like a typical California Girl. Lots of "likes" and "totallys." She had to give a testimony in a church. It was a ten-minute testimony. So as an experiment, I'd see how influenced she was by the American culture… by the *California* culture. I decided I would count the number of times she said the word "like"… you know, like, like like totally.

---

10 Ezekiel 18:18
11 Romans 6:23
12 John 3:16
13 Deaths per day worldwide: 153,558 (U.S. Census Bureau, 2004)

In ten minutes, she said "like" 81 times, believe it or not. Eighty one times, like. It really shows that you and I are subconsciously influenced by the culture that surrounds us. We're influenced by what we say, by what we even eat, by the clothes we wear, how we talk, etc. In fact I could go to a university, I could point out certain people and by the clothes they're wearing, and I could tell you what music they like, even their life's philosophy.

The Bible says we're like sheep.[14] Sheep tend to imitate one another. I come from a country that has a lot of sheep. Seventy million sheep, three million people. A lot of sheep. I can tell you, people are like sheep. Sheep tend to imitate one another. One sheep in the line jumps, the next four, five, six will jump. And I know people are like sheep because I give out a lot of Christian literature, and I know from walking along a line, and one person says, "Yeah, I'll take one," the next six people say, "Yeah, I'll have one too." One person says, "Nah…no, thanks," and the next six people will say, "No, thanks." We are like sheep.

The Bible even says we're like sheep that are prepared for the slaughter.[15] I don't know if you realize this today, but that's so true. You and I are like sheep waiting to be slaughtered. Now one thing when it comes to sheep, when they want to slaughter them…they have what's called a "Judas" sheep. Because sheep tend to follow one another and imitate one another, the Judas sheep goes to the top of the ramp and turns to the right…the rest of the sheep go *forward* and get their throats cut.

I want to show you that you're like the sheep that go to the slaughter, and what I'm going to do, if I may, is take you through the Ten Commandments and show you what a dangerous state you are in. So, please bear with me.

The First of the Ten Commandments says, "I am the Lord your God; you shall have no other gods before Me."[16] That means, God *commands*, as your Creator, the giver of life to you, that He be the focal point of your affections. You and I are *commanded* to love the Lord our God—our Creator—with all our heart, mind, soul, and strength, and to love our neighbor as much as we love ourselves.[17] Do you love the God who gave you life? Do you realize the eyes you see with, the brain you think with, the taste buds you eat with, the ears you hear with, come as a gift from God? You didn't evolve. You were created by the genius of God's handiwork.

God has blessed you with this abundant land. You're not lying in some place in Africa on the dirt with flies crawling across your face. God has blessed you beyond your wildest dreams. Do you love God? It's only right that you do. Before you say yes, let me ask you another question: Have you ever used God's name in vain? You say, "Ah, yeah, yeah. I say that GD, JC just slips out." So what you've done is taken the name of the God who gave you life and used it as a cussword to express disgust. Adolph Hitler's name wasn't despised enough to use as a cussword, but you've taken the name of God and used it in blasphemy, which is a terribly serious sin in God's sight. The Bible says, "The

---

14 Isaiah 53:6
15 Jeremiah 12:12
16 Exodus 20:3
17 Matthew 22:37–39

Lord will not hold him guiltless who takes His name in vain."[18] So if you've ever used God's name in vain, then you *don't* love God.

You love the god of your own creation, the god that is a figment of your imagination; the god that you've created to conform to yourself and your own sins. That's called idolatry when you make a god with your hands or your mind, and the Bible warns idolaters will not inherit the Kingdom of God.[19] The God who gave us life is holy and just and righteous. The Bible says He has "prepared a day in which He will judge the world in righteousness,"[20] and the standard with which He'll judge the world is the Ten Commandments.[21] Jesus said, "You've heard it said by them of old, 'You shall not commit adultery,'" but then He said, "But I say to you whoever looks upon a woman to lust after her has committed adultery already with her in his heart."[22] What will you do on the Day of Judgment when you stand before God and He judges you by *that* standard? Ever lied or stolen? Then you are a lying thief, and the Bible says, "All liars have their part in the lake of fire."[23] No thief will inherit the Kingdom of God. Can you see that you're waiting as a sheep to the slaughter, and if you die in your sins and God gives you justice, you'll end up in Hell? Folks, that's not God's will. He provided a way for you to be forgiven. And He did it through His Son, Jesus Christ. God became a perfect human being and gave His life on the cross of Calvary, taking the punishment for the sin of the world.

When Jesus was on the cross He was bruised for our iniquities. The Bible says, "For God so loved the world that He gave His only begotten Son that whoever believes in Him should not perish but have everlasting life."[24] Folks, that's what God offers you today through the gospel: everlasting life. So come to your senses. Do what the Bible says. Repent.[25] Turn from your sins and put your trust in the Savior, Jesus Christ. And the moment you do that, you will pass from death into life. You'll pass from darkness into light. Now Mr. Blower is on his way to blow me away,[26] so please think seriously about what you've heard today. Get a Bible. Don't only just pray. Get right with God. Pick up that Bible and obey what you read. The Bible is God's will for you on this earth. Thank you very much for listening this morning. God bless you.

* * *

Evil Roy Slade

A number of years ago I saw a movie called *Evil Roy Slade*. You may have seen it yourself. The opening scene is a wagon train being attacked by vicious Indians. Almost everybody is slaughtered by the Indians (it's a comedy). Then there's a cry from a crib and an Indian goes up to

---

18 Exodus 20:7
19 1 Corinthians 6:9
20 Acts 17:31
21 Romans 2:12; James 2:12
22 Matthew 5:27,28
23 Revelation 21:8
24 John 3:16
25 Acts 17:30
26 The courthouse gardeners would often drown out the preaching with their "blowers."

the crib. He's dressed in horrific war paint. He pulls back his tomahawk to kill the little baby. As he looks in, he takes a look at the face of this baby—it's Evil Roy Slade. He looks horrified and runs off in terror. The Indian had gazed on the face of Evil Roy Slade as a baby.

The scene changes and you see the baby walking through the desert. He's about two years old. He's wearing diapers; he's carrying his teddy bear and he's kicking cactuses. Evil Roy Slade is a two-year-old.

Next scene, twenty years later, you see a six-foot bronze figure, hairy chest…a man fully grown, walking through the desert, still wearing diapers, still holding a teddy bear and still kicking cactuses. Because this is Evil Roy Slade, and what the producers wanted to tell us is that, if you kicked cactuses with bare feet…you're tough, because you're the one that gets hurt. Think of it. Kicking cactuses with bare feet, you're going to get hurt. But not Evil Roy Slade.

Jesus said the same thing to Saul of Tarsus on the Road to Damascus. Saul was a man who was anti-Christian. He was running around "killing Christians for God." If you read in the Bible, he thought he was doing God a favor by torturing and killing Christians.

The Bible tells us Saul was on the Road to Damascus to kill some more Christians for God, when suddenly a light shone from Heaven, above the brightness of the sun. And a voice came from Heaven, "Saul, Saul, why are you persecuting Me?"[27]

And Saul said, "Who is it, Lord?" which is a good response—when a voice comes to you from the heavens you've got to know it's the Lord. It's got to be God.

He said, "Who is it, Lord?" What is Your name? And the voice came back, "I am Jesus, whom you are persecuting. It is hard for you to kick against the pricks."[28] That's the Old English version. In other words, "Saul, by doing what you're doing, you're kicking cactuses with bare feet." And the same applies for you and me. If we fight against God's perfect will for our lives, we're not going to win. We're kicking cactuses with bare feet. We might as well try to fight a hurricane with our fists, or a tornado by blowing at it, or try to hold back forked lightening. You *cannot* fight against God.

You say, "What do you mean, 'Fight against God'? I believe in God; I love God. He's my friend. I talk to Him regularly." You know, when I became a Christian thirty something years ago, I got the shock of my life because I saw the Bible tells us that God is not our friend. In fact the Scriptures say God is our enemy. You say, "Ah, man, I don't believe that." Well, it's what the Bible says, and the Bible is God's Word.

Writing to Christians, Paul said, "You who were once enemies in your mind by wicked works."[29] The Scriptures say that whoever's a friend of the world is an enemy of God.[30] If you think God is

---

27 Acts 9:4
28 Acts 9:5
29 Colossians 1:21
30 James 4:4

your friend it's probably because you don't understand your relationship between you and your Creator. You see, we're not ordinary people standing before God. The Bible tells us we're criminals standing before a very good and righteous Judge. You say, "I'm not a criminal; I'm a good person." Well, let's see. Let me read the Law to you. Let me go through what's called the Moral Law—the Ten Commandments—and it will help you see your state before your Creator.

Let me ask you a question. Have you kept the First of the Ten Commandments? Is God first in your life? Do you love Him with heart, mind, soul, and strength? God lavished His goodness upon you. He gave you life. You were not an accident. You didn't crawl out of some slime and evolve. God gave you a brain to think with, eyes to see with, and taste buds to enjoy good food. He gave you ears to enjoy good music. He's *lavished* His goodness upon you. You're not on some famined land, lying in the dirt with flies crawling across your face, as we often see in Africa—countries like that. God has blessed this land, and it's only right that you and I love the God who gave us life. You say, "Oh, yeah, I love God. I believe in God."

Well, listen as we cross to the Third of the Ten Commandments: "You shall not take the name of the Lord your God in vain."[31] Ever taken the name of God in vain? You say, "Ah, yeah. I do it all the time. It's just a bit of a habit." What you're doing is taking the name of God—the name that godly Jews won't even speak, it is so holy—and bringing it down to the level of a four-letter filth word to express disgust. Adolph Hitler's name wasn't despised enough to use as a cussword, and yet you've done that. And the Bible says that the Lord will not hold him guiltless who takes His name in vain.[32] Jesus warned, "Every idle word that a man speaks, he'll give an account thereof on the Day of Judgment."[33] So you haven't loved God. The Bible says we hate God without cause.[34] What you've done is violated the Second of the Ten Commandments, which says, "You shall not make yourself a graven image."[35] That is, you should not make a god, an idol, something you bow down to. And you can do it with your mind, when you create your own concept of what you believe God is like.

I did that before I was a Christian. I had this image of what God was like. He was loving and kind and a benevolent figure, and you went to Him when you had a problem. But the Bible says that God is not like that. He is perfect. He is holy. He is righteous, and the Scriptures say He'll "by no means clear the guilty,"[36] because He is just. That means He will punish injustice.

Look across to the Seventh Commandment: "You shall not commit adultery."[37] Jesus said, if you look upon a woman to lust after her, you commit adultery already with her in your heart.[38] The Sixth Commandment says, "You shall not kill,"[39] but the Bible says, "If you hate your brother you are a murderer."[40] Jesus warned if you get "angry without cause you are in danger of judgment."[41] Have

---

31 Exodus 20:7
32 Ibid
33 Matthew 12:36
34 See John 15:25.
35 Exodus 20:4
36 Exodus 34:7
37 Exodus 20:14
38 See Matthew 5:27,28.
39 Exodus 20:13
40 1 John 3:15
41 Matthew 5:22

you ever lied, or ever stolen? If you say, "Yeah, I have. Then you're a lying thief, and the Bible says, "All liars will have their part in the lake of fire."[42] No thief, no adulterer, no fornicator, no blasphemer will enter heaven.[43] The Bible makes that clear. Can you see how that Law shows us our state before a holy God? We are criminals. We've a multitude of sins, and the Scriptures say because God is good and just, "His wrath abides upon us."[44]

We're like a very heinous criminal standing before a good judge, and the judge is angry that this man has raped a young woman and slit her throat. The judge is fuming, ready to bring down that gavel. That's what God is like. His wrath abides upon us.[45] But in the Bible, it tells us that this same judge is rich in mercy. The same righteous, holy God is love and mercy, and He provided a way for your case to be dismissed. He paid your fine in the life's blood of His Son. Jesus Christ was God in human form, giving His perfect life as a sacrifice for the sin of the world. The Bible says, "For God so loved the world that He gave His only begotten Son, that whoever believes in Him should not perish but have everlasting life."[46] "God commended His love toward us in that while we were yet sinners Christ died for us."[47] Can you understand that? What a wonderful thing God has done for you. He took the punishment for your sins, so God could forgive you completely.

What you must do to partake in that forgiveness of God, that mercy He's extended toward humanity, is to repent. Turn from all sin. Confess and forsake sin, and put your trust in Jesus Christ. Don't just *believe* in Jesus. *Trust* in Him. You trust in many other things. When you go in a plane, you trust your life to the pilots. When you step in a taxi, you trust your life to the driver. When you have surgery, you trust your life to a surgeon. How much more can you trust a faithful Creator?

The Bible says, "It's impossible for God to lie."[48] You can trust *every* word that God has said in the Bible. So, please, come to your senses today. Realize that if you die today and God gave you justice, you'd end up in Hell, and that's not God's will. He's "not willing that any should perish, but that all should come to repentance."[49] God sets before you a way of life and a way of death. Choose life and live.[50] Cry out, "God, forgive me. I'm a sinner." Confess your sins to God. Forsake your sins. Put your trust in Jesus and you will pass from death to life. You'll pass from darkness into light. Every day, 150,000 people die.[51] So get right with God today. The Bible says, "Today is the day of salvation."[52]

* * *

---

42 Revelation 21:8
43 See 1 Corinthians 6:9,10.
44 John 3:36
45 See John 3:36.
46 John 3:16
47 Romans 5:8
48 Hebrews 6:18
49 2 Peter 3:9
50 See Deuteronomy 30:19.
51 Deaths per day worldwide: 153,558 (U.S. Census Bureau, 2004)
52 See 2 Corinthians 6:2.

Falling Boxes

Recently I was reminded about physical law or a law of physics that goes something like this.[53] Every action has an equal and opposite reaction. By that I mean if you push against a wall, a law of physics says that that wall pushes back to you with the same intensity that you push against the wall.

I was up a ladder when I was reminded of this law. Eight feet up the ladder. I had to pick up a piece of wood that weighed about 30 pounds. As I pushed the wood to the back of the shelf, this law kicked into action and saw myself and the ladder as one entity.

Suddenly that entity passed through me through the ladder, and the bottom of the ladder kicked out about 12 inches. So I found myself precariously hanging from the shelf about eight-feet in the air above a ladder that was about to fall. I thought, *Oh, I could get injured any second!*

I learned something about myself. I learned that I'm a very proud person, because I thought, *You know, about now I should call out, 'Help! Somebody help me!'* But I didn't want to. The reason for that is that my staff would come running in and see me hanging from an eight-foot shelf and think, *This looks stupid. What an idiot this guy is.*[54]

So I looked around, and on my right hand, on my right side was a steel cabinet. And I thought, *If I reach out from the shelf, I can put my foot on the cabinet to save myself.* And that's what I did.

We will not call out for help if we think we can save our- selves, because we're proud creatures. And why should we anyway, if we can save ourselves? Now the Bible says, "Whoever calls upon the name of the Lord shall be saved."[55] And if you've not called on the name of the Lord it's because you think you can save yourself. You say, "What do you mean, 'Save yourself'? What are you talking about?"

You may not realize it, but you are hanging in a very precarious position. When you think about it, you and I are hanging from this life over eternity. Every one of us. And in an instant we can fall through death. In the back of your mind you're thinking, "Yeah, if that happens…if there's a God… if there's a Judgment Day…if there's a Heaven or Hell, I'll be fine. My hope is that I'm a good person."[56]

---

53 It is important to realize that these messages were preached to a ready crowd. I didn't have to try to "draw" them to listen to me, as I was forced to do for so many years. To learn how to *draw* a crowd and hold them, see "Open-Air Preaching: Four-in-One." In this DVD you will see actual open-air preaching in New York, Paris, London, Santa Monica, New Zealand, and other parts of the world. Also, see *The Evidence Bible* for further teaching on open-air preaching, and *Whitefield Gold* (Bridge-Logos Publishers). All are available at www.livingwaters.com.

54 Use personal anecdotes. These are of interest to the average listener, especially if they are slightly humorous, such as me hanging from an eight-foot shelf, and being too proud to call for help. Another advantage of recounting some- thing that happened to you is that it is easier for you to recall the experience.

55 Acts 2:21

56 Proverbs 20:6: "Most men will proclaim each his own goodness…"

That's how you think you can save yourself.[57]

What I'd like to do this morning is show you that what you're hoping in is actually a parachute filled with holes. What you're doing is *extremely* dangerous. What I want to do is show you there is no steel shelf or anything for you to put your foot on to save yourself.

So, you are trusting your own righteousness[58] to save you on the Day of Judgment—and there's going to be a Day of Judgment—and you think that you are basically a good person. Let's see if you are. Let's look at the Commandments and see God's standard for morality.[59] The first of the Ten Commandments says, "You shall have no other gods before Me."[60] Is God first in your affections? Do you love the God who gave you life? You and I should. He gave us taste buds to enjoy good food. He's lavished His goodness upon us. We've a free land to be in, and enjoy the pleasures of this life—good food and friendships, health, a mind to think with, ears to hear with, eyes to see with. You're not some person in Africa, part of some famine, lying on the ground with flies crawling cross your face. God has greatly blessed you as a person. So you should love God and put Him first in your affections. Do you do that?

Some of you might say, "I do that. I love God." Well, let me ask you a question: Have you ever used God's name in vain?[61] You say, "Yeah. Once or twice." So you've taken the name of the God who lavished His goodness upon you, gave you life, and used it as a cussword to express disgust. Folks, that's called "blasphemy," when you use God's name as a cussword.[62]

Adolph Hitler's name wasn't despised enough to use as a cussword, and you've taken the name of the God who gave you life, and blasphemed Him—which shows that you *don't* love God.[63] And

---

57 Your job (with the help of God) is to dash all hope, and the biblical way to do that is with the Moral Law. You must strip your hearers of every ounce of self-righteousness, by pounding out the Law. Put them up the river Niagara without a paddle, and let them hear the roar of the falls. Then they will take hold of the rope that is being thrown to them in Jesus Christ. They will never take hold of the Savior while they are allowed to believe they can save themselves.

58 Romans 10:3: "For they being ignorant of God's righteousness, and seeking to establish their own righteousness, have not submitted to the righteousness of God."

59 This is what Jesus did: "Now as He was going out on the road, one came running, knelt before Him, and asked Him, 'Good Teacher, what shall I do that I may inherit eternal life?'" So Jesus said to him, "Why do you call Me good? No one is good but One, that is, God. You know the commandments: 'Do not commit adultery,' 'Do not murder,' 'Do not steal,' 'Do not bear false witness,' 'Do not defraud,' 'Honor your father and your mother'" (Mark 10:17–19). This is also what the apostle Paul did. See Romans 2:21–24.

60 Exodus 20:3

61 Exodus 20:7

62 Never be afraid to make sin personal. Remember what Nathan did with David. The king had sinned against God, but he deceived himself by perhaps thinking that God didn't require an account of what he had done. But look at how Nathan personalized the king's sin: "Then Nathan said to David, 'You are the man!' Then he said, 'Why have you despised the commandment of the Lord, to do evil in His sight? You have killed Uriah the Hittite with the sword; you have taken his wife to be your wife, and have killed him with the sword of the people of Ammon.'" Nathan told David that he was *personally* responsible before God. He then pointed to the Commandments, and spoke of judgment, *before* he brought the good news (the gospel)—that God had put away his sin. That is the biblical order for your message. We must make our hearers understand that they have sinned against God *before* we give them the good news of the gospel. Look at David's reaction to Nathan's reproof: "Then David said to Nathan, 'I have sinned against the Lord.' And Nathan said to David, 'The Lord also has put away your sin; you shall not die.'" (Read the entire account in 2 Samuel 12:7–13.)

63 When speaking of open-air preaching, R. A. Torrey said, "Don't be soft. One of these nice, namby-pamby, sentimental sort of fellows in an open-air meeting the crowd cannot and will not stand. The temptation to throw a brick or a rotten apple at him is perfectly irresistible, and one can hardly blame the crowd."

the problem is this: You've probably created a god to suit yourself. The Bible says there is one God and He is perfect,[64] holy,[65] and righteous.[66] The Bible says He'll "by no means clear the guilty."[67] What you've done is transgressed the Ten Commandments, which say, "You shall not make yourself a graven image."[68] That means you shouldn't make a god to suit yourself, either with your hands or with your mind.

Come across to the Sixth Commandment.[69] It says, "You shall not kill."[70] You say, "Well, I haven't killed anybody." The Bible says if you get angry without cause you're in danger of judgment.[71] The Scriptures say, "Whoever hates his brother *is* a murderer."[72] The Seventh Commandment is, "You shall not commit adultery."[73] Maybe you've not committed adultery, but listen to what Jesus said. He said, "But I say to you, whoever looks upon a woman to lust after her has committed adultery already with her in his heart."[74]

Have you ever lied or stolen? The Bible says, "No liar will inherit the kingdom of God, no thief..."[75] In fact the Bible says, "All liars will have their part in the lake[76] of fire."[77] See, if you are trusting in your own goodness to save you, realize you're not good. You're like the rest of us. Your conscience is defiled,[78] the Bible says. It even says you've got an *evil* conscience.[79] What you do when you judge yourself is judge by a standard you don't even judge other people with.[80] That's because your conscience is dead. You know it's wrong to lie and steal and lust and blaspheme. So, on the Day of Judgment you'll be guilty and the Bible says you'll end up in Hell. That's not God's will.

Folks, the same Judge that condemns us the Bible says is "rich in mercy."[81] Even though we're guilty, He's provided a way for you to be forgiven. He's provided a way of escape. He's provided a way so that your case can be dismissed.

---

64 Matthew 5:48

65 1 Peter 1:16

66 1 John 2:29

67 Exodus 34:7

68 Exodus 20:4

69 Always keep in mind that the conscience of the sinner will bear witness with the Commandments. In his heart he knows that what you are saying is true. Romans 2:15 says, "...which show the work of the law written in their hearts, their conscience also bearing witness, and their thoughts the mean while accusing or else excusing one another." Just after Paul says this, he uses the Law to bring the knowledge of sin (verses 21–24).

70 Exodus 20:13

71 Matthew 5:22

72 1 John 3:15

73 Exodus 20:14

74 Matthew 5:27,28

75 1 Corinthians 6:9,10

76 You may feel uncomfortable saying this. But the way to overcome the fear of man is to be overcome with the fear of God. Memorize 2 Corinthians 2:17: "For we are not as many, which corrupt the word of God: but as of sincerity, but as of God, in the sight of God speak we in Christ." Preach with the understanding that you do so "in the sight of God."

77 Revelation 20:8

78 Titus 1:15

79 Hebrews 10:22

80 "And do you think this, O man, you who judge those practicing such things, and doing the same, that you will escape the judgment of God?" (Romans 2:3)

81 Ephesians 2:4

The way He did it was in becoming a human being in Jesus Christ, and giving His life as a sacrifice for the sin of the world. Jesus paid your fine in His life's blood, so you could leave the courtroom on the Day of Judgment. The Bible says that "God so loved the world that He gave His only begotten Son that whoever believes in Him should not perish but have everlasting life."[82]

Folks, God offers you the gift of everlasting life. Please, today come to your senses.[83] Repent[84] and put your trust in Jesus Christ, and the Bible says you'll pass from death to life.[85] Thank you so much for listening. God bless you.

* * *

How Could I Have Missed?

A number of years ago there was a movie, a cowboy movie, starring John Wayne. John Wayne was on horseback, and a woman was lying on a rock holding a rifle. She was going to kill him because he had killed her kid brother. As he rides passed, *Blam!*[86] He hits the dirt! She smiles, walks up, and kicks his body.

Suddenly he gets his arm and hits her in the back of the legs, she hits the dirt, and he stands above her and she looks at this great hulk of a man and says, "How could I have missed?" And then

---

82 John 3:16

83 This is what happened to the prodigal son—he *came to himself:* "And when he came to himself, he said, 'How many hired servants of my father's have bread enough and to spare, and I perish with hunger!'" (Luke 15:17). The unsaved world is "insane" until it receives "the spirit of a sound mind." It has been well said that it runs at Hell as though it were Heaven and rejects Heaven as though it were Hell. In a sense, it is suicidal. We, with God's help, are to try to reason with them not to self-destruct.

84 As you preach the gospel, divorce yourself from the thought that you are merely seeking "decisions for Christ." What we should be seeking is repentance within the heart. This is the purpose of the Law—to bring the knowledge of sin. How can a man repent if he doesn't know what sin is? If there is no repentance, there is no salvation. Jesus said, "Unless you repent, you shall all likewise perish" (Luke 13:3). "God is not willing that any should perish, but that all should come to repentance" (2 Peter 3:9).

It seems that many don't understand that the salvation of a soul is not a resolution to change a way of life, but "repentance toward God, and faith toward our Lord Jesus Christ" (Acts 20:21). The modern concept of success in evangelism is to relate how many people were "saved" (that is, how many prayed the "sinner's prayer"). This produces a "no decisions, no success" mentality. This shouldn't be, because Christians who seek decisions in evangelism become discouraged after a time of witnessing if "no one came to the Lord." The Bible tells us that as we sow the good seed of the gospel, one sows and another reaps. If you faithfully sow the seed, someone will reap. If you reap, it is because someone has sown in the past, but it is God who causes the seed to grow. If His hand is not on the person you are leading in a prayer of committal, if there is no God-given repentance, then you will end up with a stillbirth on your hands, and that is nothing to rejoice about. We should measure our success by how faithfully we sowed the seed. In that way, we will avoid becoming discouraged. Billy Graham said, "If you have not repented, you will not see the inside of the Kingdom of God."

85 1 John 3:14

86 Do not whisper or scream at your hearers. If you whisper, no one will hear, and there is therefore no purpose in preaching. Even if you speak up without shouting, some may say that you are speaking too loud. These people just don't want you to preach. If this happens, answer them in a normal quiet voice to illustrate why you are lifting up your volume. More than likely they will not be able to hear what you are saying, and your point is therefore made.

John Wayne, in typical John Wayne fashion, takes his big hand, cups blood from his shoulder, smears it across her jacket and says, "Ya didn't." Typical John Wayne. Tough.

Well, that's Hollywood; but what about real life? Well, in real life—while in Hollywood John Wayne is the guy that's still shooting Indians with 200 arrows in his back—in real life, the way he died was completely different. When he was in the hospital dying of cancer, he called for Billy Graham to come to his bedside to pray with him.

You see, there's a Hollywood world and there's a real world. The Hollywood part of you says, "Well, when your number's up, it's up. When you're dead, you're dead. I'm not scared of dying." But the *real* you says, "*Oh, I don't want to die!*"

You may remember some time ago, a young Korean guy in his early 20s was taken by terrorists. They cut off his head. But just before they did, they put him in front of a video camera and let him plead for his life, and that young man looked at the camera and said, "*Oh, I don't want to die!*" You and I identified with that because we're human beings. We too have something within us that says, "I don't want to die!" That's our God-given will to live.

And that's what I want to appeal to, just for a moment this morning—your will to live. I'm asking you to set aside that proud egotistical Hollywood part of you that puts up that front, and that you humble yourself today and soften your heart because if there's one chance in a million the Bible's right when it says "Jesus Christ . . . has abolished death,"[87] you owe it to your good sense just to listen.

The Bible tells us there was a certain man whose ground brought forth plentifully.[88] That is, he had a bumper crop. He said, "Man, what'll I do?" He said, "I'll tell you what I'll do. I'll build bigger barns and stack the crops in my barns, then I'll lay back and say, 'Ah, well, I'm going to eat, drink, and be merry.'" And the Bible says God said to that man, "You fool! Tonight your soul is required of you."[89] And then Jesus said, "So is he who is rich toward himself and is not rich toward God." What does that mean: "rich toward God"? Does God want your money? Well, no, despite what televangelists tell you. God doesn't need your money. He doesn't want your money. When the Bible speaks of being rich toward God, it's speaking of something different. You see, when you're going to court, if you've transgressed the law, you are in debt to the law. You're in debt. The law demands payment, whether it be your going to prison or a monetary payment.[90]

---

87 2 Timothy 1:10

88 See Luke 12:16.

89 Luke 12:20

90 By the way, if someone asks something from the crowd that sounds like a sincere question, don't worry if you can't answer it. Just say, "I can't answer that, but I'll try to get the answer for you if you really want to know." With Bible "difficulties," I regularly fall back on the powerful statement of Mark Twain: "Most people are bothered by those passages of Scripture they don't understand, but for me I have always noticed that the passages that bother me are those I do understand." Sometimes people ask questions that are designed to hinder the gospel. The Bible warns us to avoid foolish questions because they start arguments (see 2 Timothy 2:23). Most of us have fallen into the trap of jumping

The Bible says you and I aren't rich toward God at all. We're actually in debt toward Him because we've broken His Law. You say, "What Law are you talking about?" Well, the Ten Commandments—the *Moral* Law. Take for instance the Tenth Commandment: "You shall not covet."[91] Do you know what that means? That means it's a command: you shall not desire anything that belongs to another person—his Mercedes, his wife, his house, whatever he's got—the Bible says don't you covet it. God says coveting is wrong because it leads to all sorts of transgressions. You covet something before you steal it. You covet a woman before you commit adultery with her.

Let's go to the adultery one: "You shall not commit adultery." But listen to what Jesus said. He said, "But I say to you, whoever looks upon a woman to lust after her has committed adultery already with her in his heart."[92]

Let's go to the stealing one: "You shall not steal."[93] Have you ever stolen anything in your life, irrespective of its value? Then you're a thief. Or, "You shall not bear false witness."[94] Have you ever lied, even one lie? Then you're a liar, and the Bible says all liars have their part in the lake of fire.[95] When you look at God's Law, it shows us we're in debt to it. In fact the Bible says we are weighed in the balance of eternal justice and we are found wanting.[96] That is, we're in debt. Payment must be made. Look at the First Commandment: "You shall have no other gods before Me."[97] Do you love the God who gave you life? Are you thankful? Are you grateful to Him? Everything you have comes from God. Every meal you eat. I went for 22 years as a non-Christian, eating my food without giving God thanks. I mean, what do you think of a little kid…you give him some candy and he just snatches it and walks off. You feel like grabbing that kid and saying, "Hey, kid, learn some manners. Learn some gratitude." And yet for 22 years as a non-Christian, that's what I was like. I used to eat meals without giving God thanks for them. Just hogging like a fat little brat.

Are you thankful with all God gave us—the liberty of this country, the abundance of food? The Bible says we're unthankful. We're ungrateful. We're rebellious by nature. Ever use God's name in vain?[98] You say, "Ah, yeah, a bit of a habit." You've taken the name of the God who gave you life, gave you eyes, ears, a brain to think with, the freedom of this land, the abundance of food and you've used His name as a cussword to express disgust. That's called "blasphemy." Can you see how we're in debt to Eternal Justice? And on the Day of Judgment the Bible says there'll be Hell to pay. That is, God's

---

at every objection to the gospel. However, these questions can often be arguments in disguise to sidetrack you from the "weightier matters of the Law." While apologetics (arguments for God's existence, creation vs. evolution, etc.) are legitimate in evangelism, they should merely be "bait," with the Law of God being the "hook" that brings the conviction of sin. Those who witness solely in the realm of apologetical argument may just get an intellectual decision rather than a repentant conversion. The sinner may come to a point of acknowledging that the Bible is the Word of God, and Jesus is Lord—but even the devil knows that. Always pull the sinner back to his responsibility before God on Judgment Day, as Jesus did in Luke 13:1–5. Sometimes, however, it does pay to listen to what someone says. If the person hollers, "That doesn't make any sense!" apologize to them for that fact, and start again, go through the Law, explain sin, judgment, the cross, repentance, and faith. Then sincerely ask, "Does that make sense now?" They will usually admit that it does, for fear you will start over again.

91 Exodus 20:17
92 Matthew 5:27,28
93 Exodus 20:15
94 Exodus 20:16
95 Revelation 21:8
96 Daniel 5:27
97 Exodus 20:3
98 Exodus 20:7

prison is awaiting you because you cannot pay your fine yourself. You cannot make yourself rich toward God. You cannot balance the scales.

But the Bible says, "God is rich in mercy,"[99] and He provided the way for you to be forgiven. He paid your fine Himself in the life's blood of His Son. God became a perfect human being in Jesus Christ, a morally perfect man. And He gave His life as a sacrifice for the world. God made Him who knew no sin to be sin for us, that we might be made righteous in the sight of God.[100] When someone repents and puts their trust in Jesus Christ, because of His suffering death and His resurrection, God not only forgives their sins but He makes them righteous. So that person suddenly becomes rich toward God. "For God so loved the world that He gave His only begotten Son that whoever believes in Him should not perish but have everlasting life."[101] God can forgive your sins and grant you everlasting life—if you repent and trust in Jesus Christ. That means, don't just confess your sins to a priest. Don't just confess your sins to God. Confess and *forsake* them.[102] Turn from all your sin.

The second thing you should do is trust in Jesus Christ.[103] Don't just *believe* in Jesus intellectually, but *entrust* yourself to Him—like you trust an elevator. You don't just *believe* in an elevator. It won't get you to the 10th floor. You must *trust* yourself to it. Do that with Jesus Christ and the moment you do that, God will forgive your sins. He'll grant you the gift of everlasting life. He'll give you a new heart with new desires. You'll be born again. Suddenly you'll become God-conscious.

You're like a blind man at the moment. Well, God will open your eyes to the glory of His creation. Instead of just seeing a tree you'll see that's the creative handiwork of the genius of Almighty God. "I once was blind but now I see," says the hymnist. That's what will happen to you the moment you repent and trust in Jesus Christ.

And, folks, please do it today. Don't wait until tomorrow. You may not have tomorrow. Tonight God may say to you, "You fool! Tonight your soul is required of you." So get right with God today. Repent, and trust in the Savior. God will forgive your sins and grant you His gift of everlasting life.

* * *

I Am the Center of the Universe

Some time ago a young man said to me, "How would you react if I told you that I thought I was the center of the universe?" I said, "Well, I'd react by thinking you're a normal human being." We all feel like that. For example, if I asked you where you thought "here" is—H-E-R-E—you'd probably say, "It's right where I'm standing." And I'd think the same thing. We all think "here" is where *we* are. In other words, *we're* the center of the universe.

Another word that's very difficult to pin down is the word "now." When do you think "now" is? You can't pinpoint "now." If you say it's *now*...no, no...that's *then*. Even the sentence I'm saying right *now* is in the past. Okay, so there's no *now* and then, but there's still the sun rising, and the sky is

99 Ephesians 2:4
100 2 Corinthians 5:21
101 John 3:16
102 See Proverbs 28:13.
103 See Acts 20:21.

still blue, and up is still up. Oh, no. No, it's not. The sky is not blue. You ask any astronaut. There's no color to it. The sun doesn't rise. The earth *turns* giving the impression to us that the sun actually rises. And there's no such thing as up. You say, "What do you mean, 'There's no such thing as up?'" Well, you tell me where "up" is. You point up. Say, "That's *up*." Well, you go to New York—move across the curvature of the earth and someone in New York thinks up is that way. Keep going around the earth—someone in Australia or New Zealand, and say, "Where's up?"—and *their* up is our *down*.

There's no such thing as "up"; there's no such thing as "down." The sky is not blue, the sun doesn't rise, there's no here and now.

So what can you be sure of in life? Well, there's an old saying that says, "Two things in life are sure: death and taxes." Uh, uh. That's not true. Plenty of people avoid taxes. Plenty of people, *millions*, avoid taxes—legally and illegally. So, what can we be sure of? Well, we can be sure of death. That's for sure. All of us die. Ten out of ten die. And the reason we die is because God has subjected the entire human race to something called "futility." What do I mean by that?

Well, everything you put your hand to in this life is transient. It's not going to last. It's not eternal. What it is is "temporal." It doesn't matter what it is, what riches you attain, what positions you obtain—they're going to be snatched from your hand. Even fame and fortune doesn't last. Take for instance a man called John Wayne—131 movies, a lot of cowboy movies… *very* famous man, a household name in the western world. But you ask this generation, "Who's John Wayne?" and they'd probably say, "Uh… isn't that an airport in California? Huh?"

We're subject to futility. The only thing that lasts is God. He's eternal. You and I are temporal. So, if you want to pass from the temporal into the eternal, what you must do is somehow get in contact with God. Well, how do you do that? Do you become religious? No, that won't help you. What you must do is have God do a divine operation within your heart, so He comes to know you. And then you can know Him.

You cannot help yourself. You're like a surgeon who needs a brain operation. He cannot operate on himself. It's impossible. What you need is God Himself to do a divine operation in your heart, and there's one thing that'll help that take place more than anything else, and that's your understanding of God's Law—that's your understanding of your state before a holy God.

So, let me go through that Law for a few moments and ask you some questions. All I need from you is an open heart. All I need from you is honesty. You've got nothing to lose. Please be honest and you've got everything to gain.

Have you ever lied or stolen in your life? You say, "Yes, I have." Then you're a lying thief. Have you ever used God's name in vain? You say, "Yeah, I have." That's blasphemy, when you use God's name as a cussword. Do you realize that Jesus said this: "Whoever looks upon a woman to lust after her has committed adultery already with her in his heart"?[104] Have you ever done that? You say,

---

104 Matthew 5:27,28

"Yeah, I do it all the time." Well, listen to how God sees you—as a lying, thieving, blasphemous, adulterer at heart.

You see, God sees your thought life. Nothing is hidden from the eyes of Almighty God, and the Bible tells us something about His character. It says this: God is just and holy. That means God will see to it that justice is done. He will not only punish murderers and rapists, but He'll punish thieves and liars and blasphemers and adulterers and fornicators. That puts you and I in big trouble. In fact the Bible says this: "We are enemies of God in our mind through wicked works,"[105] that God's wrath abides upon us.[106] We're like a magnetized anvil, waiting for a hammer to come down. And that hammer is Eternal Justice. Did you realize that? Do you realize that God's Word promises that all liars will have their part in the lake of fire?[107] Do you realize no thief will inherit the kingdom of God, no adulterer, no fornicator?[108] That's someone who's had sex out of marriage. In other words, you're in big trouble, and if you die in the state you're in, you have God's promise you'll end up in Hell. Now, that's not an easy thing to say. You don't make friends by saying that, but folks, it's the truth!

If you got in a car and I knew the brakes were faulty, and you were heading toward a thousand-foot cliff and I knew you were going to go over the edge, I would have to tell you. And I *have* to tell you today that God is just and holy, and you aren't getting away with a thing! No murderer commits the perfect crime. No one is going to escape justice, neither is a thief or a liar or a fornicator or blasphemer. God will bring every work to judgment, including every secret thing whether it is good or whether it is evil.[109]

So what will you do on the Day of Judgment? Folks, I don't want you to end up in Hell. *You* don't want to go to Hell! And that's not God's will. God puts before you a way of life and a way of death, and He says, "Choose life."[110] And the reason God can give you life today is because He's rich in mercy. And He sent His Son, born of a virgin—Jesus of Nazareth. He was a unique individual. The Bible calls Him "the only begotten Son of God."[111] And the reason He was unique is because He was without sin, and He was able to give His life as a perfect sacrifice for the sin of the world.

The Bible says, "For God so loved the world that He gave His only begotten Son that whoever believes in Him should not perish but have everlasting life."[112] We violated the Law of Eternal Justice, and God's anger was against each of us. But the moment we came to Christ, the Bible says our sins were washed away. Now, I'm speaking of Christians. If you're not a Christian, then you haven't experienced having your sins washed away, and you need to do what the Bible says and repent and trust in Him who died for you and rose again on the third day. "God commended His love toward us in that while we were yet sinners Christ died for us."[113] Is it nothing to you that God became a person and gave His life to take the punishment for your sins? Is your heart so hard that you can allow someone to die for your sin, and have it mean nothing to you? Folks, do what the Bible says: repent and

---

105 Colossians 1:21
106 John 3:36
107 Revelation 21:8
108 See 1 Corinthians 6:9,10.
109 Ecclesiastes 12:14
110 Deuteronomy 30:19
111 John 3:18
112 John 3:16
113 Romans 5:8

trust in the Savior. Turn from all sin and put your faith in Jesus Christ, and you'll pass from death to life. You have God's promise on it.

Something in you says, "I don't want to die!" Something in you says, "Oh, this life is futile!" Minute by minute you're getting closer to death, but the moment you repent and trust in Jesus Christ —listen to what the Bible says—"You will know the truth and the truth will make you free."[114] You will know *reality*. You'll come out of this futile transient existence, and your feet will be established on a rock, and you'll find yourself in a relationship with Him that is eternal.

Folks, please do it today. Please put your trust in Christ today. You may not have tomorrow. Every day 150,000 people pass into death, and you may not see the sun set tonight. So call upon the name of the Lord. Put your trust in the Savior and the Bible says you'll "pass from death to life."[115]

In a moment, my friend's going to come and briefly speak to you. We both want you to know we don't get paid for this. We don't particularly like doing this. It's very awkward. We don't want your money. We're not saying join a church. We're only here because we care about you and where you spend eternity. So, folks, please give it some serious thought. After he's spoken, we'll offer you a free CD called "What Hollywood Believes." Please take it—you'll not only find it fascinating, but you'll find it helpful.

⋆ ⋆ ⋆

Jesus Christ Superstar

A number of years ago, I was doing this sort of thing [preaching open-air] in a different location, in a different country, when the cast of "Jesus Christ Superstar" showed up where I was speaking. Those of you who don't know what "Jesus Christ Superstar" is, or was, it's a rock musical from the 1970s and '80s that was very popular, and the cast showed up where I was speaking.

Pontius Pilate in full garb. Caiaphas the high priest, with his "phylacteries"—dressed as the Sanhedrin dress with their black robes, great big high hat, and holding onto a staff. And I had the very strange experience of being heckled by Pontius Pilate himself as I preached the gospel. He said things like, "We rubbed you guys out two thousand years ago and you're still going!"

It was a very strange experience, and unbeknown to me, the cast of "Superstar" (who were in plain clothes) got among the crowd to whom I was speaking, and every now and then they would burst into song in perfect harmony: "Jesus Christ, Jesus Christ, who are You? What have You sacrificed?"

Afterward, they come up to me and said, "Would you come to the show this afternoon?" because they came to the local square to promote it. I said, "No. There's no way." I don't know if you know about it, but "Jesus Christ Superstar" is anti-Christian. It's actually Jesus through the eyes of Judas Iscariot. And I said, "There's no way you'll get me to come to it." Then they said, "We'll give you free tickets." I said, "What time does it start?"

So I went, and I must admit to you, I thoroughly enjoyed the music. It was enjoyable. Afterward, they invited me down the back behind the scenes to meet those who played in the play, or the actors,

114 John 8:32
115 John 5:24

or whatever. I met Mary Magdalene, the guy who played Jesus, and I said, "I really enjoyed it, but it wasn't the Jesus of the Bible that you were portraying." And then one of the members of the staff (or whatever) said, "But we're making Jesus acceptable to the twentieth century." She really nailed it.

They had made Jesus acceptable in the 20th century. Now, the Bible calls that "idolatry"—where you create a god in your own image. And you may have your image of God today. I mean, who is God to you? Well, He's the one you pray to when you want to win a football game. He's the one you pray to when Aunt Martha's sick, when things aren't going right—because God's your friend. Right?

Well, not according to the Bible. The Bible actually says something very radical. It says God is our *enemy*. You say, "You're kidding?" No, that's what the Bible teaches. It says we're enemies of God in our mind through wicked works.[116] It says whoever's a friend of the world is an enemy of God.[117] And there's nothing that dissipates idolatry (that is, creating god in your own image) like the Ten Commandments.

When God was giving Moses the Ten Commandments on Mount Sinai, the Bible tells us that Israel was worshiping a god that they made in their own image. They created a god to suit themselves, commonly called the "golden calf." They were having an orgy—running around naked and having a big party, having a good time. And Moses came down from the mountain, and the Bible says what he did is he took the two tablets and threw them at Israel's feet, smashed them to a thousand pieces.

Then he said, "Who is on the Lord's side?"[118] And Israel had to make a choice. And I'd like to do exactly the same with you today, if you'd permit me. I would like to cast those Commandments down at your feet and you make a judgment as to whether or not you've shattered them in a thousand pieces, and then call you to a decision: "Who is on the Lord's side?"

So let's do it very briefly. The Bible says, "You shall not commit adultery."[119] But listen to what Jesus said. He said, "But I say to you whoever looks upon a woman to lust after her has committed adultery already with her in his heart."[120] Broken that Commandment? I mean, who hasn't? What guy in America hasn't got "red blood"? We're guilty of lust, and the Bible says when we lust after a woman we commit adultery with her in our heart.

"You shall not bear false witness."[121] Have you ever told a lie? I'm not saying a white lie, black lie. Have you ever borne false witness? Have you ever told a boldfaced, outright lie to cover your tracks? Who hasn't? The Bible says, "All liars will have their part in the lake of fire."[122] Are you guilty? Have

---

116 See Colossians 1:21.
117 James 4:4
118 Exodus 32:26
119 Exodus 20:14
120 Matthew 5:27-28
121 Exodus 20:16
122 Revelation 21:8

you broken that Commandment? "You shall not steal."[123] If I open your wallet, your purse, and I just take out one dollar, I'm as guilty of theft as if I took out one hundred dollars. God isn't impressed with the amount of what you stole, or the value of that which you've stolen. Have you stolen something in your life? Are you innocent or guilty? Have you broken that Commandment? Want some more?

Ever used God's name in vain? The Bible says, "You shall not take the name of the Lord your God in vain, for the Lord will not hold him guiltless who takes His name in vain."[124] Have you ever used God's name in that way, without respect? When you do so—when you say, "Oh my G-D!" or "JC!"—you're taking the name of the God who gave you life itself, your most precious possession—the God who gave you eyes, a brain to think with, ears to hear with, taste buds to enjoy good food—you're taking the name of the God who lavished His goodness upon you, and using it as a cussword to express disgust.

Adolph Hitler's name was not despised enough to use as a cussword, but you've used God's name, and Jesus said, "Every idle word a man speaks, he'll give an account thereof on the Day of Judgment."[125] Can you see how we've all broken those Commandments? Can you see how we're guilty? What will you do on the Day of Judgment when God brings out every secret thing?[126] Because the Bible says it *will* happen. God will have His Day of Justice.

You think how man craves for justice. With all our evil ways, we still set up court systems. When some guy rapes a little girl, strangles her to death, throws the body in a trash heap, we all rise up and say, "That person must be brought to justice!" because we believe in right and wrong. Well, how much more will God? The Bible says that God has set aside "a day in which He will judge the world in righteousness."[127] He *will* have His Day of Justice. How will you and I fare on that day? The Scriptures say, "It is a fearful thing to fall into the hands of the living God."[128] Hell is God's just place of punishment. Murderers will be punished; so will rapists, thieves, liars, fornicators, blasphemers. What will you do on that Day?

"Jesus Christ, Jesus Christ, who are You? What have You sacrificed?" That's what they sang from the crowd. Well, who was Jesus Christ? The Bible says He was God in human form. Jesus said amazing things. He said, "He that has seen me has seen the Father."[129] The Bible says He was the "express image"[130] of the invisible God.[131] God became a person specifically to give His life as a sacrifice for the sin of the world. He paid your fine in His life's blood. He took the punishment for the sin of the world. Upon Him was laid the sin of us all.[132] The Bible says He was bruised for our iniquities.[133]

That's what God did for you and I. "For God so loved the world that He gave His only begotten Son that whoever believes in Him should not perish but have everlasting life."[134] Folks, through the

---

123 Exodus 20:15
124 Exodus 20:7
125 Matthew 12:36
126 Ecclesiastes 12:14
127 Acts 17:31
128 Hebrews 10:31
129 John 14:9
130 Hebrews 1:3
131 Colossians 1:15
132 Isaiah 53:6
133 Isaiah 53:5
134 John 3:16

suffering death and the resurrection of Jesus Christ, God can remit...He can forgive...He can wash away your sins. "Though your sins be as scarlet," the Bible says, "they shall be as white as snow."[135] You can be clean if you repent—an old-fashioned word that means to turn from your sins—and if you'll trust in Jesus Christ as Lord and Savior. Will you do that today?

Who is on the Lord's side? Who is going to say, "God, forgive me. I'm a sinner. I turn from all sin and trust you, Jesus Christ, as Lord and Savior"? The moment you do that, you pass from death into life. You pass from darkness into light. This isn't my word. I give you God's word on it. He is faithful who promised.[136] Whoever calls upon the name of the Lord shall be saved.[137]

* * *

John Wayne's Real Name

Some of you may have heard of a guy named John Wayne. If you haven't, you're probably from another planet. John Wayne acted or starred in 131 movies. We even named an airport after him locally. John Wayne, when he originally arrived in Hollywood, had the name Marion Morrison, believe it or not. There're certain things that make a Hollywood star successful, and one is you must have a name that people will remember, some sort of catchy name. Another thing you must have is a good speaking voice. A number of stars in the silent movies lost their jobs when talkies came along. And another thing you must have, if you want to be a successful star, is the ability to act.

Enter Arnold Schwarzenegger. His accent was so thick you could hardly figure what he was trying to say. The guy couldn't act. People could hardly say his name, spell his name, or even remember his name. But Arnold had something Hollywood wanted, and that was *muscle*.

I'm an author. I've been writing books for about twenty-five years, and I know if you begin writing a book or someone begins reading your book, what you must do is grip their attention on the first page. If you don't do that, you're going to lose your reader.

Enter the New Testament. If you open the New Testament, the first page, what you see is genealogies. There's a whole stack of names, about fifty names, many of which are hardly pronounceable. But the New Testament has something that you and I want. And that is, it tells us how we can find everlasting life. That's the message of the New Testament.

Some people say, "The Bible's full of mistakes." I say to them, "Okay, can you tell me what is the thread of continuity through the Bible—Old Testament through the New Testament?" People often don't know. Perhaps you don't know. But the Old Testament is God's promise that He would deliver man from death. The New Testament tells us how He did it.

135 Isaiah 1:18
136 Hebrews 10:23
137 Romans 10:13

The New Testament and Old Testament tell us that we die because we've sinned against God. We've violated an eternal Law. That Law is the Ten Commandments. And the Law is kind of like a mirror. When you and I got up this morning, one of the first things we did was go to the mirror. We did that because we wanted to see what damage had been done during the night. We look in the mirror and it's a horrible sight. The mirror tells us we need cleansing. We go from the mirror to the water to wash.

The Ten Commandments, or God's Law, is like a mirror. When we look into it, all it does is reveal what we are in truth. It reveals that we're unclean in the sight of God[138] and we need cleansing. That's all it does. So what I'm going do is hold up the mirror for you this morning. It's not going to be a pretty sight. Some of you will zone out. You'll say, "I don't want to look in the mirror. It makes me feel guilty." But you must do it, because you have to face a holy, righteous, and perfect God on the Day of Judgment, and if you face Him in your sins, there will be terrible and eternal repercussions. So, please, don't zone out. Open your heart—just listen for a few moments; you've got nothing to lose. I'm not here for your money. There's no collection bag going along. All I want you to do is open your heart and listen for a few moments.

Now, let's look at those Commandments and see if we are criminals in the sight of God. "You shall have no other gods before Me"[139] is the First of the Ten Commandments. That means you and I should love the God who gave us life, with all our heart, mind, soul, and strength.[140] I'm going to use the word "should"—the Bible calls it a *command*. We are *commanded* to love God because He gave His life; because He gave us eyes to see with, ears to hear with, a brain to think with, taste buds to enjoy good food with. God has lavished life upon you. He has blessed you with the ability to choose. I mean, that's your greatest gift from God. You can say no to the God who gave you life. There are repercussions, but it's your choice. You're not a robot. You have free will.

Have you loved God with heart, mind, soul, and strength? Before you say yes, I have a question about the Third Commandment: "You shall not take the name of the Lord your God in vain."[141] Ever done that? You say, "Oh, yeah, yeah. Yeah, I have done that." What you've done in doing that is taken the name of God—which godly Jews won't even speak or write because they consider it to be so holy—you've taken His name and used it as a cussword to express disgust. That's called "blasphemy," and the Bible says the Lord will not hold him guiltless who takes His name in vain.[142]

Jesus warned that every idle word a man speaks, he'll give an account thereof in the Day of Judgment.[143] So your very mouth—your *heart*—shows that you don't love God. In fact, you hold God and His name in contempt, and what you've probably done is what I did before I was a Christian. I violated the Second of the Ten Commandments, which says, "You should not make yourself a graven image."[144] I didn't make an idol and bow down to it, but I made an idol in my mind and bowed down to it. That is, I created a god in my image.

---

138 See Isaiah 64:6.
139 Exodus 20:3
140 Matthew 22:37,38
141 Exodus 20:7
142 Exodus 20:7
143 Matthew 12:36
144 Exodus 20:4

I had this image of God as being benevolent, and kind and loving, with no sense of justice, or righteousness or truth. And I snuggled up to this god that I'd created because I felt comfortable praying to him. And I did—every night for ten years I rattled off the Lord's Prayer to my god, who didn't exist because he was a figment of my imagination. When you make a god to suit yourself, that's called idolatry, and the Bible warns idolaters will not inherit the kingdom of God.[145]

Now listen to the Seventh Commandment: "You shall not commit adultery."[146] Maybe you haven't done that, but listen to what Jesus said. He said, "But I say to you whoever looks upon a woman to lust after her has committed adultery already with her in his heart."[147] When I read that thirty-three years ago, it was like a dart hit me in the chest. I thought, "Man, if that's the standard God's going judge with on the Day of Judgment, I'm guilty a thousand times over . . . daily. I'll be without excuse. If God considers lust to be adultery, then I'm guilty."

Have you ever stolen anything? Then you're a thief. Ever told a lie? Then you're a liar, and the Bible warns, "All liars have their part in the lake of fire."[148] Ever hated someone? The Bible says that you are a murderer.[149] That's how high God's standard is. And that's the standard He's going to judge you with on the Day of Judgment. He sees your thought life. Shall not He who formed the eye also see?[150] Shall not He who created the ear also hear?[151] That's what the Bible says. And shall not God, who gave man knowledge, also know?[152] Nothing is hidden from the eyes of Him with whom we have to deal.[153] There's not one hair on your head that God didn't make. There's not a hair on your head that He's not familiar with, the Bible teaches.[154] All things lie open and naked before His holy eyes.[155] What will you do on the Day of Judgment?

Now, can you see how the mirror shows us that we all need cleansing? The Bible says we're all as an unclean thing, and all our righteous deeds are as filthy rags in His sight.[156] You say, "What shall I do to escape the damnation of Hell? Should I clean up my life?" That won't help you. All you'll do is be a reformed criminal—a criminal who's in debt to the Law. The fact that you don't commit crimes any longer won't help you. Should you get religious? Religion can't help you. Religion is man striving to get right with God by his own efforts. What you need is mercy from God, and it's available in Jesus Christ alone.[157]

God sent His Son, born of a virgin.[158] God became a human being in Jesus Christ,[159] and this perfect sinless man gave His life as a sacrifice for the sin of the world. Jesus paid your fine in His life's

---

145 1 Corinthians 6:9
146 Exodus 20:14
147 Matthew 5:27,28
148 Revelation 21:8
149 1 John 3:15
150 Psalm 94:9
151 Psalm 94:9
152 Psalm 94:10
153 See Hebrews 4:13.
154 See Matthew 10:30.
155 Hebrews 4:13
156 Isaiah 64:6
157 Charles Spurgeon said, "Preach Christ or nothing: don't dispute or discuss except with your eye on the cross."
158 Matthew 1:23
159 See 1 Timothy 3:16.

blood, because without the shedding of blood, the Bible says there's no forgiveness of sin.[160] "For God so loved the world that He gave His only begotten Son, that whoever believes in Him should not perish but have everlasting life."[161] You say, "Oh, I believe in Jesus." Folks, it's more than *believing* in Jesus. The Bible says, "Demons believe and tremble."[162] When the Bible uses that word "belief," it means as an implicit *trust*, like you trust a parachute to save you. You don't *believe* in the parachute, you put your *trust* into it. And the Bible says, "Put on the Lord Jesus Christ."[163]

Jesus Christ, the Messiah, the Son of God, suffered and died for your sins, rose again on the third day and defeated death. And the Bible promises if you repent and trust in Him, God will forgive your sins. He'll dismiss your case. He will cleanse you by the blood of Jesus as though you never sinned, so that when you stand before God on the Day of Judgment, God can allow you to go from the demands of Eternal Justice. He can let you live.

Isn't that what you want? Isn't there a cry in your heart that says, "Oh, I don't want to die!" It's a God-given will to live. Please listen to it today. Yield your life to Jesus Christ. Cry out, "God, forgive me. I'm a sinner." Folks, He'll make you a new person. You'll be born again with a new heart and new desires. And, please, do it today. Death could seize upon you today. You just don't know what's going to happen tomorrow, and the Bible says if you say, "I do," you're a fool.[164]

⋆ ⋆ ⋆

Mexican Eye Peckers

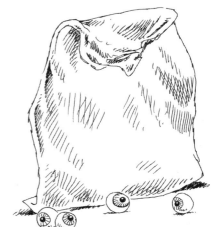

Many years ago, I had a bird aviary. If you know anything about birds, when you purchase them from bird stores, what they do is they don't put them in a cage when you buy new birds. They actually put them in a paper bag. They seal the top of the bag, they put staples in the top, and they put little holes for the birds to breath fresh air.

These two birds were called, I think, "Java Sparrows." Java Sparrows have got very fine gray feathers, bright pink eyes, red circles around the eyes and a very bright reddish beak—very impressive looking birds.

I took this bag of birds back to my office and put the bag on my desk in the office. I was very intrigued and fascinated by how people reacted when they came into my office. They'd come in and say, "Hey Ray...," and they'd talk to me about something, and suddenly they'd be distracted by this bag that was on my desk that moved by itself. They'd say, "Oh, what's that?" I'd say, "Take a look—a couple of birds."

---

160 Hebrews 9:22
161 John 3:16
162 James 2:19
163 Romans 13:14
164 See Luke 12:20.

They'd pick them up and look through the little hole, and everyone had the same reaction. They'd go, "Oh, *what sort of birds are those?*" and I took great delight in saying, "Mexican Eye-peckers." Everybody reacted exactly the same: they'd pull it back, because we want to protect our eyes.

I don't know if you realize it, but you've been created with sensors on your skin. If you lean on a hotplate, and you wait for your brain to think about it, and say, "I'm leaning on a hot plate, I could get burned!" by the time you think about it, you'll be severely burned. There are sensors on your skin that react automatically, irrespective of your will. It's an instinct. You'll pull back, and the same with your eyes. If I came along the line today and went to poke out one of your eyes, you'd immediately, instinctively, react to protect your eyes with your hands.

Now with that thought in mind, I've got a thought for you. Imagine if I was an eye doctor, and today I'm offering a million dollars for an eye. We want to transplant eyes. I could take one of your eyes out in one hour, a painless operation. I'll take the eye out and slip a little glass eye in the slot. The glass eye will look as good as the other eye, just won't *look* as good as the other eye—you'll only be able to see through *one* eye. Would you sell an eye for a million dollars?

You say, "Ah, I would consider that." Okay. Here it is: Would you sell *both* of your eyes for twenty million dollars? You say, "Hey, twenty million dollars. Whoa, *now I can see the world!*" Nope. You won't see a thing! You'll never gaze into the eyes of your children. You'll never look at a sunset again. You'll never see the blueness of the sky. You'll never look at God's creation again. You won't even see the inside of your eyelids. You'll just have blackness forever, until the day you die.

I guarantee, nobody in his right mind would *even for a moment* consider selling both of his eyes for twenty million, fifty million, a hundred million dollars. Your eyes are precious—they're priceless. And yet they're merely the windows of your soul. Your life looks through these little windows called "eyes." You've got shutters. You shut the shutters and you can't see out of the eyes. Jesus said you are to despise the value of those precious eyes, compared to the value of your soul. He put it this way: "If your eye offends you, pluck it out and cast it from you."[165] He said if your eye causes you to sin, just pull it out, priceless though it may be—and don't leave it hanging around. *Cast* it from you. And He said, "It is better for you to enter Heaven without an eye, than to go to Hell with both of your eyes." In other words, of all of the things you should prioritize in this life, it's not your health; it's not your eyesight. *It's the salvation of your soul*—that very life that's precious to you.

The Bible says, "What shall it profit a man if he gains the whole world and loses his own soul?"[166] And I want you to know why I'm here today. It's not to get your money. There's no collection. It's not to say to join a church. It's just to ask you to consider the salvation of your soul.[167] Think of your eye and how you value it. How much more you should value your eternal salvation—where you're going to spend eternity?

The way I'm going to (hopefully) cause you to consider the salvation of your soul is to just briefly go through the Ten Commandments, that you might see yourself from God's perspective.

---

165 Matthew 18:9

166 Mark 8:36

167 It is always good to remind your hearers of your motives. You don't want their money. You are not telling them to join a church. You are there because you care about them and their eternal welfare. We are to commend ourselves to their conscience: "Therefore seeing we have this ministry, as we have received mercy, we faint not; but have renounced the hidden things of dishonesty, not walking in craftiness, nor handling the word of God deceitfully; but by manifestation of the truth commending ourselves to every man's conscience in the sight of God" (2 Corinthians 4:1,2).

As you notice in our city today, there's no smog. Look around. Clean as a whistle. Blue sky, clean air. It's the *other* cities that have a smog problem. You can see the smog in the other cities. Trouble is, you go to those other cities and look around, you'll say, "There's no smog problem here." What you must do to see the truth is get in a plane and fly over Los Angeles and you see that we're breathing in black poison. When you're in the middle of it, you don't see it clearly.

I want to give you *God's* perspective on sin, because when you're a person living in sin, you don't see yourself in truth. God's perspective is this: The Bible says, "There is none righteous, no, not one."[168] It says, "We're all as an unclean thing and all our righteous deeds are as filthy rags."[169] So let's go through those Commandments so you might see yourself from God's perspective.

Let's go to the heavy one: "You shall not kill."[170] That means you shall not murder. It doesn't mean you shouldn't step on an ant. It means you shouldn't take the life of another human being. You say, "Ah, well I haven't committed that sin." Listen to what the Bible says: "Whoever hates his brother is a murderer."[171] The Bible says God requires truth in the inward parts.[172] He considers the desire is the same as the deed.[173]

"You shall not commit adultery."[174] But listen to what Jesus said: "Whoever looks upon a woman to lust after her has committed adultery already with her in his heart."[175] Ever hated someone? Ever lusted after another person? Then you've violated those two Commandments.

Ever told a lie? Ever stolen something? Then you're a lying thief. Ever desired something that belonged to somebody else? That's "covetousness"—a violation of the Tenth Commandment. The First Commandment says, "You shall have no other gods before Me."[176] That means God should be first in your affections. He lavished His goodness upon you. He gave you life. He gave you those eyes that you value, that brain you think with, those hands with which you pick up things. *God* created you. You're not an accident.

The Bible says He's familiar with every hair upon your head—every thought within your heart. The Psalmist said, "There's not a word on my tongue, but lo, O Lord you know it altogether."[177] Nothing you've ever said God hasn't heard. Nothing you've ever done God hasn't seen. He sees *everything*. All things lie open and naked before the eyes of Him to whom we have to give an account.[178]

Now think of those Commandments. Think of how you've used His name in vain, used the name of God as a cussword. Think of how you've violated the Second Commandment by creating a god in your own image—making a god to suit yourself, when there is one God who is perfect, holy and righteous. And He's the one you've got to stand in front of.

Folks, I know this isn't pleasant—looking at our sins from God's perspective, though we have to face Him on Judgment Day. And if we die in our sins, the Bible says we'll end up in Hell. The Scrip-

---

168 Romans 3:10
169 Isaiah 64:6
170 Exodus 20:13
171 1 John 3:15
172 Isaiah 51:6
173 See Romans 7:7–14.
174 Exodus 20:13
175 Matthew 5:27,28
176 Exodus 20:3
177 Psalm 139:4
178 Hebrews 4:13

tures warn, "All liars have their part in the lake of fire."[179] And the Scriptures say the same judge, the same holy, righteous Creator is rich in mercy.[180] It's because of His mercy He became a person in Jesus Christ. Jesus came to this earth to suffer and die for the sin of the world. The Bible says, "He was bruised for our iniquities."[181] Jesus paid your fine in His life's blood so you could leave the courtroom.

You're waiting to go to court today, probably because you've violated a law. Imagine if someone paid your fine. You accept that payment, then you're free to go. The judge will say, "You're out of here; the fine is paid." That's what Jesus Christ did for you in His life's blood. He suffered and died for the sin of the world. "God commended His love toward us in that while we were yet sinners, Christ died for us."[182]

While we were rebels, holding up fists in rebellion to God—unthankful, unholy, not concerned about the God who gave us life, not considering His demands for one moment, drinking in iniquity like water—"while we were *sinners* Christ died for us," and then He rose again on the third day and defeated death. And the Bible says, "The wages of sin is death but the gift of God is eternal life through Jesus Christ our Lord."[183] He is the only one who can save you from God's wrath. "There is no other name under heaven given among men, whereby you must be saved"[184]—only Jesus Christ. So please consider seriously the claims of the Bible. Think about your eternal salvation, because, again, there is *nothing* more important than where you spend eternity. Remember what Jesus said: "What shall it profit a man if he gains the whole world and loses his own soul?"[185]

So realize today that "God commands all men everywhere to repent, because He has appointed a day in which He'll judge the world in righteousness."[186] So make sure you repent. Don't just *confess* your sins to God. Confess and *forsake* them. And don't just *believe* in Jesus; *trust* in Him as you trust a parachute to save you.

Folks, *please do it today*. You may not be here tomorrow. Did you know that every single month in America alone, 45,000 people die of cancer? Cancer could be within your body at the moment. The odds are one or two people in this line have got terminal cancer. You say, "I hope it's not me." It's always someone else. Human nature believes that. But if you found that you had cancer today, you would cry out, "Oh, God, *I don't want to die.*" That's your God-given will to live. Listen to it, because you *have* a terminal disease. You *are* dying. Today you could be swept into eternity. Could be tomorrow somebody finds they've got cancer, at least they are stopped in their tracks. At least they're humbled by the disease and turn to God, so they have an advantage over you who are not concerned with the fact that you're going to die—that you are terminal. So think about these issues and think about why I'm speaking to you. This isn't fun. I don't like doing this. I don't want your money; I'm not saying join a church. I do it only because I care about you and where you spend eternity. So please give that some consideration.

---

179 Revelation 21:8
180 Ephesians 2:4
181 Isaiah 53:5
182 Romans 5:8
183 Romans 6:23
184 Acts 4:12
185 Matthew 16:26
186 Acts 17:30,31

* * *

No Strings Attached

Anumber of years ago, I was driving my car and I saw a package on the road. I got real excited because I saw a free package, stopped the car, backed it up, jumped out, . . . and couldn't find the thing. So I began to look *under* the car, to see if it was *under* the car. As I did so, I saw a piece of string running from a ditch into a second story window and realized I'd been fooled by the same trick that I used as a kid to taunt motorists. What you do is put a package on the road, put string to it, hide in the bushes, and when a motorist stops you pull the package in and watch him make an idiot of himself looking for it.[187]

You get something in your mailbox. It's free—like a little free sachet of coffee or perhaps toothpaste. It isn't really free. There are *strings* attached to it. Anything you get from this world often has strings attached. You get nothing for nothing—they want to attach strings to your wallet.

If you look in the Bible in the Book of Romans, it says the gift of God is eternal life, but it adds the word "free" before the word "gift," which makes it seem superfluous because when something is free it's a gift; and if it's a gift, it's free.

The reason that word "free" and "gift" are together, or those words are together is to substantiate something very, very important. And that is, there is no way you can earn everlasting life from God. It *must* be a free gift. And if I walk down this line, I guarantee most of you would think that you're going to *earn* everlasting life. Because I'd say something like this to you, "Would you consider yourself to be a good person?" You say, "Yeah." "Do you think you'd make it to Heaven? Why?" "Well, because I'm basically a good person." The reason you'd say that is because you don't see your true state before God.

The Bible actually tells us we're not ordinary people as we stand before God. We are *criminals*, in debt to His law. And that's why our goodness won't get us out of the state that we're in.

Let me put it this way. Think of a man who has committed rape and murder and is standing in front of the judge. He says, "Judge, I know I'm guilty but the lady was a whore; I mean she *deserved* it, so I'm sure you'll just let me go." His statement makes it obvious that he hasn't seen the serious nature of his crimes. So what the judge must do is read the law to him and show the seriousness of his crime by the judgment the law is going to give to him. He'll say, "You violated the law. You've murdered a human being. You're going to the electric chair!" That might bring the seriousness as to what he's done, to the criminal's heart.

---

187 Make sure that if you bring humor into your message that it isn't sarcasm or the type of humor that humiliates. A good type of humor to use is to relate something you did that was dumb. A laugh from the crowd can be regretted when a few minutes later a sincere person openly accuses you of humiliating someone in the crowd. No matter how nasty your heckler, you will find it difficult to justify yourself in the light of the admonition for us to be "blameless."

So what I'm going to do today is read the Law to you and then show you the judgment that's coming by the Law, so you'll see the seriousness of your crime against God, so that you'll see that you cannot earn His favor. So, let's look at God's Law just for a few moments.

The First Commandment is, "You shall have no other gods before Me."[188] That means God, the One who gave you life, commands that He be the focal point of your affections. You and I are commanded to love the Lord our God with all our heart, mind, soul, and strength and to love our neighbor as much as we love ourselves.[189] Do you love the God who gave you life? Do you realize where your life came from? Do you realize where your eyes came from, your mind, your ears, and your taste buds? Isn't it true, God has lavished His goodness upon you?

You have eyes to see this creation of God. You've ears to hear good music. You've taste buds to enjoy good food. God has lavished His goodness upon this land. You're not in Africa lying in the dirt with flies crawling across your face. God has blessed you. Do you love Him because of that? Are you thankful to Him? Before you say, "Yes, I am. I believe in God. I love God. I'm thankful to Him," if you used His name as a cussword, if you've taken His name in vain, that's called "blasphemy"—where you substitute the name of the God who gave you life for a four-letter filth word to express disgust. The Bible says, "The Lord will not hold him guiltless who takes His name in vain."[190] And the fact that you used His name in vain shows you do not love God with your heart, mind, soul, and strength. You couldn't care less about Him.

What you've probably done is transgress the Second of the Ten Commandments, which says, "You shall not make yourself a graven image."[191] You love your *image* of God. You're like me for 22 years before I was a Christian. I had a god of my image. I prayed to a god of my own creation—a god who didn't create Hell, a god who wasn't just and holy and righteous, a god who was a figment of my imagination, the place of imagery—and then I bowed down to the non-existent god that I created, and I felt comfortable, snuggling up to him. And yet there is one God and He's perfect and holy and righteous. And the Bible says, "He'll by no means clear the guilty."[192]

You go through those Commandments—like the Sixth Commandment: "You shall not kill."[193] The Bible says if you hate someone, you're a murderer.[194] Jesus said if we get angry without cause we're in danger of judgment.[195] Seventh Commandment: "You shall not commit adultery,"[196] but listen to what Jesus said of that Commandment. He said, "But I say to you whoever looks upon a woman to lust after her has committed adultery already in his heart."[197] That leaves every guy guilty, plenty of women guilty. If you as much as lust, you commit adultery in God's eyes.

Ever stolen something, irrespective of its value? Ever lied? You say, "Yeah, I've done that." Then you're a lying thief. And the judgment by the Law is death and damnation, eternal Hell: "All liars have their part in the lake of fire."[198] The Bible makes it very clear.

---

188 Exodus 20:3
189 Matthew 22:37
190 Exodus 20:7
191 Exodus 20:4
192 Exodus 34:7
193 Exodus 20:13
194 1 John 3:15
195 Matthew 5:22
196 Exodus 20:14
197 Matthew 5:27,28
198 Revelation 21:8

Can you see the serious nature of your crimes against God? Can you see that you haven't got anything to bargain with Him? When you stand before Him on Judgment Day, you're not going to say, "God, You have to let me live because I'm good."

You've got no "goodness" to bargain with. The Bible says our righteous deeds are as filthy rags in His sight.[199] We're criminals. All we can do is fling ourselves on the mercy of the Judge and say, "God, be merciful to me a sinner."[200] The Bible says, "God is rich in mercy to all that call upon Him."[201] And the reason that God can show you mercy is because He paid your fine 2,000 years ago. God became a perfect human being in Jesus Christ. And this perfect man gave His life as a sacrifice for the sin of the world. He paid the fine in His life's blood. He was bruised for our iniquities.[202] The Bible says, "For God so loved the world that He gave His only begotten Son, that whoever believes in Him should not perish but have everlasting life."[203] And then He rose from the dead on the third day after He suffered for our sins, defeating the grave. And all who repent and trust in Him receive remission of sins. God dismisses their case. He allows us to live because of what Jesus did on the cross.

Folks, do you understand that?[204] Because multitudes are going to stand before God for all those crimes against them, and say, "God, I'm basically a good person." Such a deception! That's why the Bible says the gift of God—the *free* gift of God—is eternal life. You *cannot* earn a gift. So today, please come to your senses. Cry out, "God, forgive me. I'm a sinner." Willfully put your trust in Jesus Christ, as you trust a parachute to save you. Exercise biblical repentance. That's more than just saying, "God, I'm sorry." It means, "God, I'm sorry, and I'll *never* do it again." That's the attitude of true repentance. You say, "I can't do that. You mean to say I have to not lust anymore?" Yeah. You don't want to run around as a hypocrite. It means if your eye causes you to sin, you as much as pluck it out and cast it from you. That's how serious sin is. Better you be blind and end up in Heaven than have your eyesight and lust and end up in Hell. Jesus said that.[205] Please do it today. Ask God to forgive you. Trust in Jesus Christ. Do it today because you may not be here tomorrow—every single day

---

199 Isaiah 64:6

200 Luke 18:13

201 Ephesians 2:4, Romans 10:12

202 Isaiah 53:5

203 John 3:16

204 It is obvious from Scripture that God requires us not only to preach to sinners, but also to *teach* them. The servant of the Lord must be "able to teach, patient, in meekness instructing" those who oppose them (2 Timothy 2:24,25). For a long while I thought I was to leap among sinners, scatter the seed, then leave. But our responsibility goes further. We are to bring the sinner to a point of understanding his need before God. Psalm 25:8 says, "Good and upright is the Lord: therefore will he teach sinners in the way." Psalm 51:13 adds, "Then will I teach transgressors your ways; and sinners shall be converted to you." The Great Commission is to teach sinners: "teach all nations...teaching them to observe all things" (Matthew 28:19,20). The disciples obeyed the command: "daily in the temple, and in every house, they ceased not to teach and preach Jesus Christ" (Acts 5:42). The "good-soil" hearer is he who "hears ...and understands" (Matthew 13:23). Philip the evangelist saw fit to ask his potential convert, the Ethiopian, "Do you understand what you are reading?"

Some preachers are like a loud gun that misses the target. It may sound effective, but if the bullet misses the target, the exercise is in vain. He may be the largest lunged, chandelier-swinging, pulpit-pounding preacher this side of the Book of Acts. He may have great teaching on faith, and everyone he touches may fall over, but if the sinner leaves the meeting failing to understand his desperate need of God's forgiveness, then the preacher has failed. He has missed the target, which is the understanding of the sinner. This is why the Law of God must be used in preaching. It is a "schoolmaster" to bring "the knowledge of sin." It teaches and instructs. A sinner will come to "know His will, and approve the things that are more excellent," if he is "instructed out of the Law" (Romans 2:18).

205 Matthew 5:29

150,000 people die.[206] They pass from time into eternity. And today you and I could die. Of course, we don't believe that. But you can guarantee the people, the 150,000 people that died yesterday didn't believe it either. So please, get right with God.

* * *

One of the Dumbest Statements

Years ago I got into a friend's vehicle and said to him, "Going to put on your seatbelt?" and he said, "Uh, I'm not planning on having an accident." I said to him, "Dave"—which was his name—I said, "Dave, that's got to be one of the dumbest statements I've ever heard in my life: *'I'm not planning on having an accident.'*"

You see, no accident is ever planned. If you don't believe me, check out the dictionary definition of accident. In fact, if you *plan* to have an accident and put "accident" on your insurance form, you'll probably end up in prison.

Back in the 1980s there was an advertisement on television with some dummies—*crash* dummies—having an accident, and this deep voice says, "What goes through the mind of a dummy if it's not wearing a seatbelt in a head-on collision?" Well, what *does* go through the mind of a dummy in a head-on collision if it's not wearing a seatbelt? Then the voice says, "The steering wheel." That's what goes through the head of a dummy in a head-on collision if he's not wearing a seatbelt. And then it said, "Don't be a dummy. Buckle up!"

That's what's commonly called "scare tactics," but nobody complained about scare tactics on an advertisement, because it was a *legitimate* scare tactic. It *is* a fearful thing to have a steering wheel go through your forehead. Who would deny that?

You see, there is a fear that's our friend. It's a fear that's for your self-preservation. It's a fear that says, "Don't step off a thousand-foot cliff." There's a fear that says, "Put on a parachute before you jump out of a plane at 25,000 feet." Fear is your friend in some cases, not your enemy.

The Bible says, "It's a fearful thing to fall into the hands of the living God."[207] There is the ultimate plea for self-preservation. Now what does it mean, "fall into the hands of the living God"? Well, if you violate civil law you "fall into the hands of the judge," and the inference is that you don't *place* yourself in the hands of the judge—you *fall* into his hands. There is something happening beyond your control, and that is, the law has put you into the hands of the judge, and your fate is in his hands.

The Bible says, "It's a fearful thing to fall into the hands of the living God." There is something that's out of your control that is going happen to you after you die. You're going to stand before the living God, and the Bible says "it's a fearful thing" to fall into His hands. Why? Because another law

---

206 Deaths per day worldwide: 153,558 (U.S. Census Bureau, 2004)
207 Hebrews 10:31

will put you there. It's called "the Law of sin and death"[208]—the Moral Law, the Law that we commonly refer to as the Ten Commandments.

So what I'm going to do is take you briefly through the Law, and I hope your fear that causes self-preservation will kick in, and you'll see that that fear is your friend and not your enemy…and that you'll do something about it.

Let's look at the Tenth Commandment: "You shall not covet."[209] Have you ever desired something that belongs to somebody else? I mean, have you desired someone's husband, wife, their car, their house? Desired to be rich? Well, you've broken the Tenth Commandment. You've been covetous. Or the Ninth Commandment: "You shall not bear false witness."[210] Have you ever told a lie? I'm not talking about a half-truth, or a fib, or an exaggeration. I'm talking about bearing false witness, sinning against your conscience—knowing the wrong and doing the wrong. Have you ever violated that Commandment? Now listen to what the Bible says: "All liars will have their part in the lake of fire."[211] Does that scare you? It scares the living daylights out of me to know that God sees lying as an abomination.[212] It's a detestable thing to tell a lie in the sight of God. And His judgment is that all liars will have their part in the lake of fire. If you've lied once, you are a liar. Listen to your fears because that's a fear for self-preservation. Or, "You shall not steal."[213] Have you ever stolen something irrespective of its value? If you've stolen, you're a thief and you cannot enter the kingdom of God.[214]

Have you ever used God's name in vain? Have you used God's name as a cussword to express disgust? Oh, that's a fearful thing. This is the God who created the lightning, the thunder; this is the God who made all things. He hears everything you've said. Nothing is hidden from His eyes, and the Bible says He will not hold him guiltless that takes His name in vain.[215] What will you do on that Day? Oh, folks, listen to your fears. That fear that's speaking to you now is your friend. Listen to it.

Or the Seventh Commandment: "You shall not commit adultery."[216] But listen to what Jesus said. He said, "But I say to you whoever looks upon a woman to lust after her has committed adultery already with her in his heart."[217] Have you ever looked with lust? Then you've committed adultery in the sight of a holy God. Have you ever hated someone? The Bible says you're a murderer.[218] Ever created a god in your own image—that is, created a god in your mind you feel comfortable with? That's called idolatry, and it's a transgression of the Second of the Ten Commandments.

Or how about the First: "You shall have no other gods before Me"?[219] Do you love God above all things? Is He the focal point of your affections? Do you realize that you were created for His pleasure?[220] Did you know that? That's the purpose of your existence: for the pleasure of God. Does your

208 Romans 8:2
209 Exodus 20:17
210 Exodus 20:16
211 Revelation 21:8
212 See Proverbs 12:22.
213 Exodus 20:15
214 See 1 Corinthians 6:9,10.
215 Exodus 20:7
216 Exodus 20:4
217 Matthew 5:27,28
218 1 John 3:15
219 Exodus 20:3
220 See Revelation 4:11.

life bring the God who created you pleasure? And if you're honest you'll say, "No. No, I've sinned against Him. I've violated His Commandments." And you realize that His wrath abides upon you, not His pleasure. And that if you died in your sins and God gives you justice, and He judges you according to your sins, you'll end up in the lake of fire—and that's a fearful thing.

It is a fearful thing to fall into the hands of the living God. The apostle Paul said, "Therefore, knowing the terror of the Lord, we persuade men."[221] Folks, will you be persuaded today? You say, to do what? Just to open your heart for a few moments and realize what God has done for you. This God who is the God of justice, the God of righteousness, the God of holiness, is a God who is rich in mercy. And the Bible says He sent forth His Son, born of a virgin. Jesus Christ was God in human form—a perfect man who came to give us life as a sacrifice for the sin of the world.

He came to lay His life down and take the punishment for the sins that you and I've committed. Oh, folks, God's love is so great for you and I that He created for Himself a body and filled that body as a hand fills the glove, for the specific purpose of suffering and dying for our sins—for paying our fine in His life's blood so that we could leave the courtroom on the Day of Judgment. So that God could extend His mercy toward us. "For God so loved the world that He gave His only begotten Son, that whoever believes in Him should not perish but have everlasting life."[222]

Folks, that's what God offers you: salvation from death and Hell. Now listen to those fears. Listen to the voice of your conscience. Realize there's something in you that's God-given and that's the little voice that says, "Oh, I don't want to die." And then do what the Bible says. Realize that God cared enough to make provision for your sins to be forgiven. He rose from the dead in Jesus Christ and defeated death. He rose from the dead on the third day. Now the grave has lost its sting, the Bible says. What you must do to partake in that gift of God's mercy, to receive His gift of everlasting life, is to repent—turn from all sin—and trust in Jesus Christ as you would trust a parachute to save you.

Don't just believe in Jesus, trust in Him implicitly for your eternal salvation. And, folks, the moment you do that you have God's promise that you'll pass from death into life. Most people, the Bible says, will not listen. Like this man here is trying his best to distract this lady from hearing the gospel. Most people will not listen. The Bible says straight is the gate and narrow is the way that leads to life and few there be that find it. Broad is the path that leads to destruction, and many go in that way.[223]

Folks, don't be deceived like most people. Cry, "God, forgive me. Open my eyes, open my heart." He'll transform you. You'll be born again with a new heart and new desires. You'll pass from darkness into light, from death into life.

* * *

Parking Tickets

A friend of mine recently told me that he had to go to court because he had a stack, a *handful* of unpaid parking tickets. I said, "Why didn't you just go and pay them?" He said, "You're kidding—they were just parking tickets. There was no big deal." Then he told me that at 4:00 a.m.

---

221 2 Corinthians 5:11
222 John 3:16
223 See Matthew 7:13,14.

the police came to his home, arrested him, and put him in a big black bus and took him to the L.A. Court system. He said, "Fortunately I took $700 in my pocket, that I thought would make things right with the judge."

As he stood before the judge he said, "Judge, I am guilty, but I want to tell you I've got money in my pocket and I can pay those fines today." The judge said to him, "Mr. So-and-So, I'm going to save you all that money. *You're going to jail.*" He brought down the gavel, and took him off to jail.

My friend said he was horrified. His big problem was that he *trivialized* his crimes. He said, "It's just parking tickets. No big deal." But the law said it was a big deal. And because of his trivializing of his crimes, he really thought that he could fix things up with the judge.

You and I have violated another law. It's the Law of God, the Ten Commandments, and I know if I came down the line today and spoke to you personally you would probably trivialize your crimes. You say, "Yeah, yeah, I have sinned. I've lied, I've stolen... but it's no big deal." And, "I've told white lies... just *little* lies. I've stolen *little* things." *You're trivializing it.* What you need to hear is what the Law says, so you can see how serious it is, so you can understand there's no way you're going to fix things up with God when you stand before Him on Judgment Day. *There's no way you can do that.* You *cannot* pay your own fine. You cannot get out of it.

So what I'm going to do is tell you what the Law says, so you can see the seriousness of your transgression and then we'll take it from there. Ninth Commandment: "You shall not bear false witness."[224] Have you ever lied? *Don't* trivialize it. Don't say to yourself, "Yeah, yeah, but they're white lies." No, face it. Just say, "Yes, I have lied, therefore I am a liar." The Bible says, "All liars have their part in the lake of fire."[225] *All* liars will have their part in the lake of fire! You say, "Well, what are you saying? I'm going to Hell because I told a lie?" *All* liars have their part in the lake of fire. You say, "That's pretty heavy." Sure it's heavy! I don't even like saying it. But that's what the Bible teaches. That's what the Law demands. We trivialized it. God says it's an abomination to bear false witness.[226] You say, "Oh, I don't believe in Hell." That's like my friend saying to the judge, "Judge, I don't believe in jails." The judge would say, "Excuse me? Take him away!" Just because we don't believe something, it doesn't go away.

Let's look at the Eighth Commandment. Ever stolen something? Don't trivialize it; face it. If you've stolen something, you're a thief, and the Bible says all thieves will end up in Hell.[227] You're a lying thief, and you have to stand before God.

Look at those other Commandments: "You shall not commit adultery."[228] You say, "I haven't done that." But listen to what Jesus said: "Whoever looks upon a woman to lust after her has commit-

---

224 Exodus 20:16
225 Revelation 21:8
226 See Proverbs 12:22.
227 1 Corinthians 6:9,10
228 Exodus 20:14

ted adultery already with her in his heart."[229] You see, God's Law "requires truth in the inward parts."[230] God is omniscient. He's omnipresent. He's been a witness to your crimes. Nothing is hidden from the eyes of Him with whom we have to do.[231] And not only that, your conscience has borne witness.[232] Isn't it true, your conscience tells you when you've done wrong? You know you've lied and stolen. Listen to that conscience.[233]

Ever used God's name in vain? Listen to the Scriptures: "You shall not take the name of the Lord your God in vain. For the Lord will not hold him guiltless who takes His name in vain."[234] All blasphemers will end up in Hell. It's taking God's holy name and using it as a cussword. Don't trivialize it, because if you do you'll be deceived into thinking you can make things right with God when you face Him on Judgment Day—you just talk God out of justice. There's no way that's going to happen.

Sixth Commandment: "You shall not kill."[235] That means you shouldn't take the life of another human being. You shouldn't murder them. But Jesus said, "If you get angry without cause you're in danger of judgment."[236] And the Bible says, "Whoever hates his brother is a murderer."[237] Can you see the trouble you're in? Can you see that if you died in your sins and God gave you justice, you'd end up in Hell? *What are you going to do on the Day of Judgment?* You haven't put God first; you haven't loved the God who gave you life. That's a *commandment*: "You shall love the Lord your God with all of your heart, mind, soul, and strength, and love your neighbor as yourself."[238]

We have trouble loving our loved ones, let alone strangers—loving them as much as we love ourselves. We've fallen short of that Law that we're commanded to keep. We're Law-breakers. We're criminals in the sight of a holy God, and the Bible says, "He that believes on the Son has everlasting life. He that believes not shall not see life, *but the wrath of God abides on Him.*"[239] What will you do on the Day of Judgment when you face that Law?

If my friend had only *known* what the law demanded, he would have attempted to make things right between him and the law. And fortunately you can do that today, because the Bible says the Judge of the Universe is rich in mercy. Almighty God is love and grace and mercy, besides being justice, righteousness, and holiness. This God sent His Son, born of a virgin; Jesus Christ was God in human form, who gave His life as a sacrifice for the sin of the world. He stepped in and paid your fine for you in His life's blood. He suffered and died for your sins, for my sin, for the sin of the whole world. "For God so loved the world that He gave His only begotten Son, that whoever believes in Him should not perish but have everlasting life."[240] The Bible says, "The Law came by Moses but grace and truth came by Jesus Christ."[241] Grace is God's unmerited favor—His mercy. And that's

---

229 Matthew 5:27,28
230 See Psalm 51:6.
231 Hebrews 4:13
232 See Romans 2:15.
233 Encourage your hearers to listen to the voice of their conscience. It is an impartial judge on the courtroom of the mind, echoing the truth of what you are saying.
234 Exodus 20:7
235 Exodus 20:13
236 Matthew 5:22
237 1 John 3:15
238 Matthew 22:37,38
239 John 3:36
240 John 3:16
241 John 1:17

what He extends to you today through the gospel. Jesus Christ suffered and died for your sins, commending God's love, *proving* God's love to you. And then He rose from the dead. The Bible says on the third day He rose from death, and defeated it. Death couldn't hold Him.

So, folks, realize today that God wants to forgive your sins. He doesn't want you to perish. He wants to grant you the gift of everlasting life. He wants to extend His mercy toward you. The Bible says that "mercy rejoices over judgment."[242] That means, God would rather give you mercy and allow you to live than give you justice—send you to Hell. That's not God's will. So come to your senses today. Do what the Bible says: "Repent." It's an old-fashioned word. It just means to turn from your sins. It means to confess and forsake your sins. It's more than confessing them to a priest or to God; it's actually *turning* from them. It means, stop being a hypocrite and saying "I believe this," and living a life that's an abomination to the God you profess to love.

The second thing you must do is *trust* in Jesus Christ. Folks, don't just *believe* in Jesus. I "believed" in Jesus before I was a Christian. If you look at a building ten stories high and you want to get to the top, you don't *believe* in the elevator…you put your faith into it, you *trust* yourself to it. And that's what you must do with the Savior. The Bible says today is the day of salvation. Now is the accepted time.[243] That means get right with God today because you may not have tomorrow.

Every single day, 150,000 people die.[244] You can be swept away in a hurricane, a flood, or your heart can give out for no reason. Eternity is a heartbeat away from you. Realize that. You're not immortal. You are not God. You are a mortal human being with frailties, and you don't know when death could seize upon you.

The only thing you can guarantee is the air going into your lungs at the present moment. You can't guarantee another breath. That comes by the mercy of God and if God loses patience with you today, folks, there's no second chance. So obey the Bible's command: "Repent." Trust in Jesus Christ, and the moment you do that, you'll pass from death into life, from darkness into light.

* * *

Spitless Natives

It's said that a very primitive tribe in Africa has a unique way of dealing out justice to guilty parties. If there's been a murder in the tribe, what they do is line up the suspects, and the guy who can't spit is put to death as the guilty party. And the reasoning is this: The human conscience is so strong—that knowledge of when we do wrong—and what happens is when you line a guy up, his conscience accuses him of his guilt; the guilt produces a fear and the fear makes his mouth go dry. So he's spitless. So he's the guy who dies for the crime. Conscience is an incredibly powerful thing.

An old drunk was once walking home, staggering on the sidewalk, when his dog came out and watched him coming toward him. The dog loved his master. The drunk walked up his pathway. The dog snuggled up behind him, and followed him into the house. The drunk collapsed on the floor,

242 James 2:13
243 See 2 Corinthians 6:2.
244 Deaths per day worldwide: 153,558 (U.S. Census Bureau, 2004)

and the dog snuggled up to his hand and waited for that moment when his master would awaken, so he could lick him and have fellowship with him. He so loved his master.

Around midnight, the dog began to bark and bark. The drunk awoke and thought, "Man, the last thing I feel like is a barking dog." He reached down and grabbed a wooden chair and threw it at the dog, and then collapsed on the floor.

In the morning when the drunk awoke, he opened his eyes and saw a sight he could hardly believe. Everything was taken from his house. During the night, thieves had broken in and stolen all

his possessions. Only two things were left in the house. One was a broken chair and the other was a dead dog.

The dog had heard thieves breaking into the drunk's home, tried to warn his master, and his master had killed his very best friend—the one that was trying to warn him that thieves were breaking into his home.

Your conscience is like that. It's like a *barking* dog. It's not pleasant when it barks at you, but what you've got to realize is your conscience is your friend, not your enemy. It's there to warn you when you do wrong, because when you do wrong there are always negative repercussions. Maybe not right now, but in the future.

What I'd like to do today with your help, your co-operation, is show you why you should be very concerned and why you should listen to that conscience. This is why: The Bible tells us that God has set aside a day in which He'll judge the world in righteousness.[245] Now that should please you if you're a good person. The day is coming when murderers will be brought to justice. What a wonderful day! Everybody who's good should rejoice when murderers are brought to justice. Oh, it'll be a day when rapists will also be brought to justice, where vicious men have viciously raped innocent women—those wicked people will be brought to justice. So good people can rejoice. Oh, it'll be a day when thieves are also brought to justice…and liars, and blasphemers and fornicators and adulterers. And good people will rejoice when that comes to pass.

Have you ever lied? You say, "Yeah, yeah, I have." Well, then, you're a liar and you're in big trouble. Have you ever stolen? You say, "Yeah, just little things." It still makes you a thief, if you've stolen something. Have you ever used God's name in vain? Have you ever used your Creator's name as a cussword to express disgust? The Bible says the Lord will not hold him guiltless who takes His name in vain. Have you ever had sex out of marriage? Then you're a fornicator. Ever committed adultery? You say, "No, I haven't done that one." Well, listen to what Jesus said: "Whoever looks upon a woman to lust after her has committed adultery already with her in his heart."[246] Have you ever done that? You say, "Have I ever done that! That's all I ever do!" Well, realize this: that Almighty God has seen every sin you've ever committed. Nothing is hidden from His eyes.[247] Nothing. "All things lie open and naked before the eyes of Him to whom we have to give an account."[248] "God will bring every

---

245 See Acts 17:31.
246 Matthew 5:27,28
247 See Hebrews 4:13.
248 Hebrews 4:13

work to judgment including every secret thing, whether it is good or whether it is evil."[249] In fact, Jesus said, "Every idle word a man speaks he'll give an account thereof in the Day of Judgment."[250] So, we're all guilty.

Okay, what is the punishment for sin? Listen to it: The Bible warns, "*All* liars will have their part in the lake of fire."[251] You say, "What? You're kidding." That's what it says: "*All* liars." Listen to your conscience. It will accuse you, tell you you're guilty…and right now you should be spitless! If your conscience is doing its duty, if that dog is barking at you, *listen to it.*

Let's go back to the African tribes. The primitive African tribe had a chief. He was a good man— but he found that there was a rumor about a mutiny within his tribe. So he called the tribe together, and he said this: "If I find anyone—*anyone*—is even *thinking* of a mutiny, I'll give them a hundred lashes publicly." A short time later, he found out that his own beloved brother had been jealous of him. He had a secret hidden jealousy within his heart, and he'd been planning a mutiny. So what he did is had his brother tied to a tree, and then he got out a whip, and he picked the strongest man from the tribe and he gave him the whip. Then the chief had *himself* tied to the tree and took the punishment for his brother. Now, in doing so, he kept his word; justice was done—a hundred lashes were given out. But he also displayed and proved his love toward his brother—that he cared for him.

Folks, that's what God did for you and I. We're guilty. We're heading for Hell, every one of us. The Bible says God's wrath abides upon us.[252] Every time we sin, we store up wrath,[253] and God says He'll give anguish to every soul that is contentious.[254] That means argumentative, rebellious, holding your fist up to God and saying, "It's my life. Don't tell me what to do!" That's rebellion. It's a sinful heart. God says that wicked soul will be punished.

Folks, if you die today, the Bible gives you a promise. It says, "The soul that sins, it shall die,"[255] and if you die in your sins you'll end up in Hell. You have God's promise for that. But just as that chief proclaimed that punishment for transgression for mutiny, so God has proclaimed punishment for transgression of His Law—eternal damnation. But He's rich in mercy, and like the chief, He took the punishment for us. God became a human being and allowed Himself to be nailed to a cross. The Bible says He was punished for the sin of the world. When John the Baptist saw Jesus for the first time, he said, "Behold, the Lamb of God who takes away the sin of the world."[256] Can you understand that?[257] That God's love for you is so great that He became a perfect human being and gave His life as a sacrifice. He paid the fine so you could leave the courtroom. He shed His life's blood and cried out in anguish, "My God, My God why have You forsaken Me?"[258]

God's wrath came upon Him, so you could go free. "He was bruised for our iniquities."[259] "For God so loved the world that He gave His only begotten Son that whoever believes in Him should not

---

249 Ecclesiastes 12:14
250 Matthew 12:36
251 Revelation 21:8
252 See John 3:36
253 Romans 2:5
254 Romans 2:8,9
255 Ezekiel 18:4
256 John 1:29
257 Your hearers must understand what happened on the cross. That is where they will see the love of God.
258 Matthew 27:46
259 Isaiah 53:5

perish, but have everlasting life."[260] God commended, displayed, *proved* His love toward us, in that while we were yet sinners Christ died for us.[261] And then Jesus rose from the dead on the third day. He *defeated* death. The Bible says, "It was not possible that death could hold Him."[262]

Now God offers everlasting life, forgiveness of sins, to all those who repent and call upon the name of Jesus Christ—all those who trust in the Savior. God says, "I'll remit your sins, I'll cancel them out, I'll blot them out,[263] I'll remove them as far as the east is from the west."[264] He says that "I'll give you right standing with Myself," and that means as a Christian, guilty though I am of breaking God's Law a multitude of times (I'm no different from any one of you), because of God's mercy, because of His grace and washing away my sins thirty-three years ago when I put my faith in Christ, that means I can stand before God boldly on the Day of Judgment[265] with no fear. I won't be spitless. I'll be filled with joy, knowing my sins are washed away, and knowing that God has granted me everlasting life—not because of my goodness, but because of God's mercy and His grace.

So, folks, obey the command of God. The Bible says, "God commands all men everywhere to repent because He has appointed a day in which He'll judge the world in righteousness."[266] Put your trust in Jesus Christ today, as you trust a parachute. Some of you say, "I can't have faith in God. I just can't see Him." Come on, when you get on a plane you trust your life to a pilot you've never seen. You don't walk into the pilot's cabin and say, "Show me your credentials. I won't trust my life to you until you do that." We don't do that. We trust men we've never seen with our lives, without thinking twice about it. So how much more should you trust a holy God? Jesus said, "Have faith in God."[267] Trust Him. He is faithful who promised.[268] Do it today because you may not have tomorrow. Every single day, multitudes, literally thousands of people are swallowed by death. Death *seizes* upon them. They plan for tomorrow and tomorrow never comes. So obey what the Scriptures say. It says, "Today if you hear His voice, don't harden your heart."[269] Today is the day of salvation. Now is the accepted time.[270]

* * *

The Fifty-Dollar Cat

A man once looked into a store, into a window, and saw a sight he could hardly believe. There was a cat sitting in the window of the store. It was a mangy cat. It was a flea-bitten, rat-ridden ugly cat. His ear was chewed off. His tail was chewed off. Half its fur was missing.

---

260 John 3:16
261 Romans 5:8
262 Acts 2:24
263 See Acts 3:19.
264 Psalm 103:112
265 See 1 John 4:17.
266 Acts 17:30
267 Mark 11:22
268 Hebrews 10:23
269 Hebrews 3:7,8
270 See 2 Corinthians 6:2.

Then he widened his eyes because next to the cat it had a sign. It said, "Cat for Sale—$50." He thought, *What? Someone wants fifty bucks for a flea-bitten, rat-ridden mangy cat?* He couldn't believe it.

Then he saw something that widened his eyes even further. The cat was drinking milk from a saucer that he recognized as being a Ming Dynasty saucer, worth thousands of dollars. He immediately grabbed his wallet, ran into the store and he said, "Hey, mister, I'll take that cat. Thanks. And here's fifty dollars." He picked up the cat and he said, "I'll just take the saucer to keep the cat company." The guy said, "You leave that saucer there. I've sold seventy cats with that saucer this week!"

That man was snared by his own craftiness, and the Bible says God has ensnared the wicked by their own craftiness.[271] What does that mean, that God has ensnared the wicked by their own craftiness? Let me explain.

Many years ago I had a kid's club, and after the club I'd give a big bag of candy to the kids. I'd just give them one each. I said, "Line up for your candy, kids," and a hundred kids lined up. As I looked at that line, I thought, *Man, look at that line. Right at the front were these little brat bullies, pushing their way to the front.* As the line went down, right at the back were these meek, quiet, sickly kids, and I thought, *Man, that is a line of greed if I ever I saw one*, and then I had an idea.

I said, "Kids, stay right where you are. Everybody turn about-face." I said, "If you move out of line you're not getting any candy." And every kid turned about-face, and I had great delight in taking that big bag of candy to the other end of the line, and giving it first to those meek, quiet, sickly kids— much to the disgust of the fat brats at the back.

Folks, that's how God has ensnared the wicked in their own craftiness. He has turned the line around. In this world where the rich get richer and the poor get stomped on, God has done something incredible. Do you know what it is? Let me tell you.

Do you believe the Bible account of Adam and Eve, Noah and the ark, Jonah and the whale, Daniel and the lion's den? You say, "Wha…Whatdya…Of course I don't believe those silly stories!" Behold the wisdom of God. The Bible says He's "chosen the foolish things" to confound the wise.[272] Why, anyone with any intellectual dignity, any social standing, would never stoop to believe those stupid Bible stories.

Folks, that's the way God has turned the line around. The Bible says He "resists the proud and gives grace to the humble."[273] And the ultimate in God's wisdom is this Bible verse. Listen to it: "God has chosen the foolishness of preaching to save them that believe."[274] I know from doing this for many years, the wisdom of God is seen on the faces of those who are too proud to even listen to the gospel, to some guy preaching to them, *because it's beneath their dignity.*

---

271 1 Corinthians 3:19
272 1 Corinthians 1:27
273 James 4:6
274 1 Corinthians 1:21

The Bible says, "Everyone who is proud of heart is an abomination to the Lord."[275] God has turned that line around, and made the door of salvation to eternal life very low. Only those who stoop in humility may enter. God resists the proud and gives grace to the humble; so I hope you'll bear that in mind because I'm going share with you the gospel of eternal salvation. I'm going to tell you how the Bible says you can have everlasting life. And I trust that you will forsake your pride, that you'll humble yourself before the mighty hand of God today.

This morning, I don't know if you heard that thunder, I *love* thunder. I love thunder that rattles the windows of my home. It doesn't scare me in the slightest. You know why? Because I know who's in control of the thunder. I know who's in charge of the weather department. It's Almighty God. When I hear thunder, it thrills my heart because I know I have peace with the God who created that thunder.

So let's go through the gospel of salvation. Do you realize that God Himself is going to judge the world in righteousness? And I'm not talking about Doomsday. I'm talking about the Day of Judgment, the day of justice when Almighty God, who is good and righteous, punishes murderers and rapists. Now if you're a good person you'll say, "Wow, that's great; there's going to be a day of justice." God is going to see to it that people who have murdered other people, who have taken the lives, the precious lives of human beings . . . God Himself is going to punish those people. That's wonderful!

You'll rejoice if you're a righteous person, but if you're a guilty criminal before God you'll cringe at the thought of Judgment Day. So let's look at God's Law: "You shall not bear false witness."[276] Have you ever told a lie? Are you innocent or guilty of breaking that Commandment? Have you ever borne false witness? You say, "Yes, I have." Then you're a liar in God's sight. Have you ever stolen something? You say, "Yeah . . . just little things." Then you're a thief, irrespective of what you've taken. Have you ever used God's name in vain? You say, "Yeah, yeah . . . a bit of a habit." Then think of that thunder! And you've taken the name of the God who created thunder and lightning—you've taken His name and used His holy name as a cussword to express disgust! The Bible says, "The Lord will not hold him guiltless who takes His name in vain."[277] Well, look at the Seventh Commandment: "You shall not commit adultery."[278] But Jesus said, "Whoever looks upon a woman to lust after her has committed adultery already with her in his heart."[279] Have you ever looked with lust? Well, God says you've committed adultery in your heart. You're in big trouble on the Day of Judgment! Have you ever had sex out of marriage? The Bible says, "Fornicators will not inherit the kingdom of God."[280]

Now, how's your conscience doing? Conscience means "with knowledge." *Con* is "with," *science* is "knowledge." Whenever you and I sin we do it with knowledge that it's wrong, because God has placed an impartial judge in the courtroom of the mind. And every time you lie or steal or blaspheme or lust or commit adultery or fornicate, there is a small voice within our mind saying: "You've done wrong! You've done wrong!"

Folks, listen to the conscience! It's not your enemy; it's your friend. It's trying to warn you! It's doing what fear does for you when you're going to step off a cliff. It says, "Uh, oh . . . don't do that, there are fearful consequences! There's a 1,000-foot cliff under your feet; pull back!" That's what your

---

275 Proverbs 16:5
276 Exodus 20:16
277 Exodus 20:7
278 Exodus 20:14
279 Matthew 5:27,28
280 See 1 Corinthians 6:9,10.

conscience should be doing. You should be listening to it today as we go through those Commandments, because the conscience will affirm the truth of each Commandment.

"You shall have no other gods before Me."[281] Is God first in your life? Do you love the One who gave you life? Do you love Him "with all your heart, mind, soul, and strength?"[282] Do you "love your neighbor as yourself"? That's what that Commandment entails. How far we fall short of the glory of God; how far we fall short of that standard—what He requires of us. And He'll require "an account of every idle word"[283] we've spoken—every deed done in darkness, every thought that's gone through our minds.

God has seen every deed done in darkness. He sees as the pure light. What will you do on that Day? The Bible says, "All liars will have their part in the lake of fire."[284] No thief will inherit the kingdom of God, no blasphemer, no fornicator, no adulterer.[285] What will you do on that Day when God manifests all of your sins, and they come out as evidence of your guilt? What a fearful thing to fall into the hands of the living God!

But the Bible tells us that "God is rich in mercy."[286] This Judge—this awesome Judge, this Creator who made every one of us—is rich in mercy, and the Bible says mercy rejoices over judgment.[287] In other words, God would far rather give you mercy than give you justice. He would far rather give you Heaven than give you Hell. He made a way for you to be forgiven. The Bible says that "God was manifest in the flesh."[288] God revealed Himself in human form in the person of Jesus of Nazareth.

God became a human being to suffer and die on the cross to take the punishment for the sin of the world. When Jesus was on the cross, it was not a good man being punished by evil men. This was Almighty God paying the fine in His life's blood, so you and I could leave the courtroom on the Day of Judgment. God "made Him who knew no sin, to be sin for us that we might be made righteous in the sight of God."[289] "For God so loved the world that He gave His only begotten Son that whoever believes in Him should not perish but have everlasting life."[290]

Folks, that's what God offers you through the suffering death and the resurrection of Jesus Christ. God offers you salvation from death. He offers you the gift of eternal life. But you must repent. Turn from all sin and trust in the Savior.

Now, earlier on I said that God has chosen the foolish things of the world to confound the wise. Listen very carefully: If you're a proud person, if you're an arrogant or self-righteous person, even the name of Jesus Christ will be an offense to you. The name of Jesus will offend you, if you're proud of heart. But the Bible tells us: God has given Him a name that's above every name, that at the name of Jesus Christ every knee shall bow…and confess Him as Lord.[291] Today if you'll not repent, in humility, and trust Jesus Christ as your Savior and your Lord, you will bow your knee to Him as your Judge.

---

281 Exodus 20:3
282 Matthew 22:37–39
283 Matthew 12:36
284 Revelation 21:8
285 See 1 Corinthians 6:9,10.
286 Ephesians 2:4
287 See James 2:13.
288 1 Timothy 3:16
289 2 Corinthians 5:21
290 John 3:16
291 See Philippians 2:9–11.

As I said, it's a fearful thing to fall into the hands of the living God. Folks, don't be foolish today! Cry out, "God, forgive me. I'm a sinner." Repent and trust Jesus Christ and you have God's promise that you'll pass from death into life. Folks, if you've got a Bible at home, please realize this: the Bible is a spiritual book, and it will not make sense to you until you're born of the spirit. It's like a little kid. Chinese is a very, very complex language, and you and I would find it very hard to pick up a book written in Chinese and read it. And yet a four-year-old can do that. A four-year-old Chinese child can do that. He can understand a book written in Chinese. Why? Because he was *born* into a Chinese family—he speaks the language. And when you're born again, when you're born into God's family, you will speak God's language, and the Bible will begin to make sense to you.

So take the time today to consider what you've heard. Repent, trust the Savior, pick up a Bible, and obey what you read. Suddenly God's Word will begin to make sense to you.

* * *

The A-Frame Roof

A number of years ago in the U.S., a man decided to paint the A-frame roof of his house. So what he did (he had a lot of ingenuity) was, he got a piece of rope and he threw it over the top of the A-frame roof, and he went around the other side and he tied it in a very firm knot to the bumper of his car. He wasn't silly—it was a good *tight* knot. Then he went around the back. He got a ladder, climbed up on the roof, secured the rope around his waist and a very firm knot (he wasn't silly). And then he leaned back and he got that paint and he began painting the roof, thrilled with his ingenuity.

A short time later, his wife—not knowing what he had done—came out with her car keys. (This is a true story.) She got in the car, started it up, and drove off, pulling him over the roof and seriously injuring him.

Now the moral of the tale is, *you're only as secure as that to which you secure yourself.* If you're the sort of person who believes that man evolved from apes, you will live your life accordingly. You'll see yourself as a glorified ape. You won't live yourself in respect to what Almighty God says. You won't believe the Bible, which says that "God made man in His own image,"[292] and "every animal brought forth after its own kind."[293] In other words, dogs don't have cats, horses don't have cows, and apes don't evolve into people. It doesn't matter how long you leave them.

As I said, you'll be as secure as that to which you secure yourself. So if you think man came from apes, you'll just live your life doing what you want as a glorified animal, fulfilling whatever your heart desires . . . and that will be to your eternal downfall.

292 Genesis 1:26
293 Genesis 1:21–25

Let me tell you what the Bible actually tells us. It says this: "God made man in His own image," and He made him so he could bring forth after his own kind. He also made man as a *moral* being—separate from the beasts. You are different from a horse, a cow, or a dog—because you're a *moral* being. How do I know that? Because you set up court systems. When you see injustice, you know justice should be done. Dogs don't care about that. You don't see "dog jails." You don't see dog judges or horse or cow judges. But you see people—human beings, because we are made in the image of God. We have a conscience. We know right from wrong. The word "conscience" means "with knowledge." *Con* is "with," *science* is "knowledge." Conscience. So when we lie, or steal, or kill, or commit adultery, we do it "with knowledge" that it's wrong.

God has given light to every man.[294] Now, because we're made in God's image, that tells us that God is something like us. He is a *moral* being. Nobody in his right mind says, "Ah, yeah…God is evil." We know God is good by nature, and because God is good we know He believes in justice; and the Bible actually tells us this. It says that God has "set aside a day in which He will judge the world in righteousness."[295] The Bible continually calls God a judge—a good judge, a righteous judge, a judge who will see that justice is done.[296]

The Bible gives us a little warning. Well, it's actually a *big* warning. It says this: that God will require "an account of every idle word we have spoken";[297] that "God will bring every work to judgment, including every secret thing, whether it is good or is evil."[298] That should thrill you if you're a good person, because good people love justice being done. When some guy slits the throat of ten young girls after he raped them, and he gets brought to justice, good people say, *"Yes! They got him!"* So it should please you that God is going to have a Day of Justice, *if you're a good person.*

That will be the day that murderers are punished, rapists are punished, thieves are punished, liars are punished, blasphemers and fornicators and adulterers.

Now, that should make you a little nervous. I'll tell you why. The Bible says each of us are guilty of those crimes. So what I'm going to go through is God's Law, just for a few moments, so you can judge yourself—whether you're innocent or guilty.

Have you ever lied? Have you ever stolen? If you say, "Yes, I have," then you're a lying thief. You cannot enter Heaven. The Bible says, "*All* liars have their part in the lake of fire."[299] Have you ever committed adultery? You say, "No. No, I haven't, although I'd like to." You see, if you lust after a woman, the Bible says you commit adultery already with her in your heart.[300] And if you have the desire, God says you're guilty of the deed. In fact, the Scriptures say if you "hate your brother, you're a murderer."[301] If you hate someone, you'd like to see them killed. That's what hatred is. That's the standard God's going to judge you and I with on the Day of Judgment.

So, will you be innocent or guilty? If you're honest, you know you'd be guilty of violating those Commandments. But add to that the fact that God has seen your thought life. Nothing is hidden

---

294 See John 1:9.
295 Acts 17:31
296 See Genesis 18:25; Psalm 96:13.
297 Matthew 12:36
298 Ecclesiastes 12:14
299 Revelation 21:8
300 Matthew 5:27,28
301 1 John 3:15

from the eyes of Him to whom we have to give an account.[302] God made your brain. He knows what you're thinking. He made the darkness, so He sees what you do in darkness as in pure light.[303]

The Bible tells us that "God is not willing that any should perish."[304] God doesn't want you to go to Hell. It's not His will, and He's proved that by sending Jesus Christ to suffer and die for our sins. God, the Judge of the Universe, is rich in mercy, and He provided a way for you and I to be forgiven, and it was through the blood of His Son. God became a perfect human being, gave His life on a cross, taking the punishment for the sin of the world. "For God so loved the world that He gave His only begotten Son that whoever believes on Him should not perish but have everlasting life."[305]

Folks, think of it. What is your greatest fear in life? Isn't it the fear of death? The Bible says you and I are tormented by the fear of death every moment of our lives.[306] Listen to your own heart. Isn't there something in you that says, "Oh, *I don't want to die*"? Isn't there something in you that says life is the most precious thing you have? The Bible says, "What shall it profit a man if he gains the *whole world* but loses his own soul?"[307] The most precious thing you have is your life, and God offers you immortality. So what should you do? Well, realize that God made provision for you to be cleansed, for you to be forgiven, through the suffering death and the resurrection of Jesus Christ. And what you must do is repent—an old-fashioned word, it means to turn from your sins. And you must put your faith in Jesus Christ. Don't just *believe* in Him intellectually; *trust* in Him implicitly for your eternal salvation. Cry, "God, forgive me. I'm a sinner," and then *trust* Jesus Christ as your Lord and Savior, like you trust a parachute to save you. You put your faith into it.

Now maybe today you say, "Look, I'm a Christian. I've given my life to Christ." A lot of people have. But you don't have the things that the Bible says "accompany salvation."[308] Well, this is what the Scriptures say: "Every tree that brings not forth good fruit will be cast down and cast into the lake of fire."[309] It'll be cut down. That means there're a lot of people who think, "Yeah, I'm right with God," but they're not, and their life proves they're not.

Basically, you're a hypocrite. You say, "I believe in God, I believe in Jesus Christ," but your life doesn't show what you know it should. So today repent, and say, "God, I've been like Judas. I've been with the disciples doing what the disciples do, but my heart has been a betrayer, in my heart." And ask God to cleanse you and forgive you and make you a new person in Christ, and then live in holi-

---

302 See Hebrews 4:13.

303 "O Lord, You have searched me and known me. You know my sitting down and my rising up; You understand my thought afar off. You comprehend my path and my lying down, and are acquainted with all my ways. For there is not a word on my tongue, but behold, O Lord, You know it altogether. You have hedged me behind and before, and laid Your hand upon me. Such knowledge is too wonderful for me; it is high, I cannot attain it. Where can I go from Your Spirit? Or where can I flee from Your presence? If I ascend into heaven, You are there; if I make my bed in Hell, behold, You are there. If I take the wings of the morning, and dwell in the uttermost parts of the sea, even there Your hand shall lead me, and Your right hand shall hold me. If I say, "Surely the darkness shall fall on me," even the night shall be light about me; indeed, the darkness shall not hide from You, but the night shines as the day; the darkness and the light are both alike to You" (Psalm 139:1–12).

304 2 Peter 3:9

305 John 3:16

306 See Hebrews 2:14,15.

307 Mark 8:36

308 Hebrews 6:9

309 Matthew 3:10

ness. Read your Bible every day because the Bible says, "Without holiness, no one will see the Lord."[310] It's been well said of the Bible, "This book will keep me from sin, and sin will keep me from this book."

You know, when someone is anti-Christian, a great preacher once said, "Follow them home, and see why." If there's something in you that's offended by what I'm saying, I guarantee I know what it is. You are doing something you know you shouldn't do in the sight of God. It's guilt. We're like cockroaches that run from the light. Have you ever gone into a dark room and turned on the light? Cockroaches scatter, and the Bible says, we "love the darkness and hate the light."[311] But the Day will come when that light will come on, and you have to stand before God. So please think seriously about what you heard today, and seek God with all your heart.

* * *

The Bible and All It Contains

A young man was once called by his lawyer and told that his grandmother had left him an inheritance. He was real excited, went to the lawyers, and the will was read to him. He found that his grandmother had left him $20,000 and "my Bible and all it contains." He thought, *Man, I want the twenty thousand bucks, but I know what the Bible's about . . . don't want that religious junk.*

So what he did was he took that Bible and put it up on a high shelf, and just left it there. He didn't open it. And then he went to Vegas and he spent the $20,000 on fast living, fast women. It was gone pretty quick.

Then he lived the next sixty years as a pauper, scraping for every meal, with barely the clothing on his back. He became so destitute, so hungry, that his relatives had to come and take him to live with them. As he was cleaning out his home, he reached up to that shelf and grabbed that old dusty Bible with its brass clasps, and as he took it in his trembling hands it fell on the floor, *and opened up to reveal a twenty dollar bill between each page.*

He had lived his life as a pauper, scraping for every meal when he could have lived his life as a rich man, but he didn't . . . because of prejudice.

Now, maybe you are prejudiced too. If you're a normal human being, you're thinking to yourself, "Yeah, I don't want that religious junk—men in nighties (night clothes), flinging water on people. I know what the Bible 'contains.'" That was kind of my philosophy . . . until I opened the Scriptures thirty-three years ago and saw that the Bible actually contains the greatest riches any of us could hope to find.

Let me ask you a question. Would you sell one of your eyes for a million dollars? If I was a doctor and I was walking along the line and saying, "Hey, we want to do eye transplants, and we can take one of your eyes out in a one-hour operation. It'll be totally painless and we'll give you another glass

310 Hebrews 12:14
311 John 3:19

eye to slip in the slot. It will look just as good as this eye. It just won't *look* as good as this eye. You'll be blind in one eye. A million dollars cash...would you take it? Some of you may.

Okay, here's the crunch. We want a matching pair. They must be fresh. We're willing to offer you twenty million dollars for *both* of your eyes. You say, "Wow! Twenty million dollars! Now I can see the world." Uh...you don't see a thing if you don't have eyes. You just sit in the darkness of your home counting your money.

I guarantee nobody in his right mind[312] for one moment would ever *consider* selling both his eyes for twenty million, a hundred million, twenty *billion* dollars. Your eyes are precious to you. Now think about it. Think of the value of your *life* that looks out of those eyes. These are the shutters that we look out of. What must your life be worth? The greatest riches you could ever hope to obtain in this life is to retain that life that you've got. Isn't that true? Jesus said, "What should it profit a man if he gains the *whole world* and loses his own soul?"[313]

So today I want to bring before you some very serious thoughts that will help you see the value of your life to you. I'm going to take you through a few of God's Commandments. Those Ten Commandments...you know, the ones that are written on your heart? "You shall not kill, you shall not steal, you shall not lie, you shall not commit adultery." They are written on the heart via the conscience. The Bible tells us God has written with the pen of a diamond, on our heart.[314] That conscience is like a judge in the courtroom of your mind. When you do something wrong, the conscience makes an impartial judgment and says, "That was wrong."[315] That's irrespective of what you want to do. You may not want to hear the voice of your conscience. It's still there. It's irrespective of what you will, and it's because God has placed it there to tell you right from wrong. So listen to that voice today as we go through a few of the Commandments. "You shall not steal."[316] Ever stolen something? If you've stolen anything, you're a thief. God saw the crime that you've committed. Have you ever borne false witness?[317] I'm not talking about white lies and half-truths. I'm talking about a boldfaced lie. Have you ever lied? You say, "Yeah, I have." Well, then, you're a liar. God has been a witness to your crime, and the Bible says, "All liars will have their part in the lake of fire."[318] I don't like saying that, but that's what the Bible says.

You and I demean sin; we trivialize it. But God sees it as something being *very* serious, worthy of the death sentence—worthy of damnation. That's how holy God is. "You shall not take the name of the Lord your God in vain."[319] Have you ever used God's name as a cussword to express disgust— something called blasphemy? Ever done that? You say, "Yeah, I have." You've taken the name of the God who gave you those eyes you so value, who gave you the ability to see and think, and hear and taste, the God who lavished you with life, and you've taken His name and used it as a cussword to express disgust. That's called "blasphemy," and the Bible says, "The Lord will not hold him guiltless

---

312 It is important to add the words "in his right mind." If someone calls out, "I would sell my eyes," you can then gently say, "I said, 'In his right mind.'"
313 Matthew 16:26
314 See Jeremiah 17:1.
315 See Romans 2:15.
316 Exodus 20:15
317 Exodus 20:16
318 Revelation 21:8
319 Exodus 20:7

who takes His name in vain."[320] What are you going to do on the Day of Judgment? Jesus said, "Every idle word a man speaks, he will give an account thereof on the Day of Judgment."[321]

Listen to this one: Jesus said, "You've heard it said by them of old, you shall not commit adultery. But I say to you, whoever looks upon a woman to lust after her has committed adultery already with her in his heart."[322] You see, God sees your thought life. Nothing is hidden from the eyes of Him with whom we have to give an account.[323]

The Bible asks the question, "Shall not He who formed the eye also see?"[324] Do you think God created the eye? Such is the genius of God's creative hand. He made the eye—do you think God is blind? No, the Bible says, "The eye of the Lord is in every place beholding the evil and the good."[325] He sees the darkness as pure light, and "He will bring every work to judgment, including every secret thing, whether it is good or whether it is evil."[326] What will you do on that Day when you stand before God and all your sins come out as evidence of your guilt? "How will you escape the damnation of Hell?"[327] That's what the Bible asks.

Well, there's an answer. God has provided a way for you to save your life. But listen to the words of Jesus: "He that seeks to save his life will lose it."[328] If you try to hold onto your soul, your life, and reject God, you're going to lose your life. But if you yield it to God through Jesus Christ, God says you can keep it.

And this is why you can keep your life. God became a human being. God was "manifest in the flesh,"[329] the Bible says, for a specific purpose. God became a perfect human being in Jesus Christ so that He could give His life as a sacrifice for the sin of the world. When Jesus was on the cross, "He was bruised for our iniquities."[330] He took the punishment for the crime that you and I committed. We broke God's Law and Jesus paid our fine in His life's blood. "He poured out his soul unto death,"[331] the Bible says. "For God so loved the world that He gave His only begotten Son that whoever believes in Him should not perish but have everlasting life."[332] Folks, that's what God offers you.[333] Jesus rose from the dead on the third day, and defeated death. The Bible says, "It was impossible for death to hold Him."[334] He defeated the grave. Now, if you repent and trust in Him as Lord and Savior, God says He will remit your sin. He'll wash away your transgressions. He will dismiss your case. He will commute your death sentence. You do not have to go to Hell. You do not have to be damned, because of God's love and mercy in Jesus Christ.

---

320 Ibid
321 Matthew 12:26
322 Matthew 5:27,28
323 Hebrews 4:13
324 Psalm 94:9
325 Proverbs 15:3
326 Ecclesiastes 12:14
327 Matthew 23:33
328 Matthew 16:25
329 1 Timothy 3:16
330 Isaiah 53:5
331 Isaiah 53:12
332 John 3:16
333 Sinners are dull of hearing, so repeat the same truth over and over from different angles. The great Puritan Richard Baxter said, "Screw the truth into men's minds."
334 Acts 2:24

So what should you do to receive this gift that God offers? Do exactly the same thing you do when you receive a gift from anybody. You turn to them and receive it. What you've got to do with God is repent—turn from sin toward God. Stop sinning. Stop lusting. Stop lying. Stop hating. Stop gossiping. Stop stealing. You say, "Man, I can't do that. Lust is such a part of my heart." Folks, when you're born again, God makes you a brand new person with a new heart and new desires.

And the second thing you must do is put your trust in Jesus Christ, in the same way you trust in an elevator. You want to get to the tenth floor? You don't "believe" in the elevator, you put your faith *into* it. And that's what you must do with the gift of God. And, folks, please do it today. Do you realize that every year 54 million people die? Every day 1,500 people die of cancer. You don't know when you're going to go. Forty-two thousand people every year in the U.S. die in car accidents. Every time you step into a car you could be stepping into a coffin.

So get right with God today. Cry out, "God, forgive me. I'm a sinner." And He will. The moment you put your faith in Jesus, you pass from death into life.

* * *

The Cat Up the Tree

In 1976 the army was brought in to help out with the British Fireman's strike. They were struck—no one to put out fires, so the army came in. They received a call (this was in 1976) from an elderly lady whose cat was stuck up a tree.

Out came the army and rescued the cat; they discharged their duty. The elderly woman was so thrilled that she invited the army into her home for tea and cookies. Afterward, fond farewells were given, and off drove the army—*over the cat, and killed it.* That's from a book called *Heroic Failures.*

Now, if you study the Christian faith, you'll find the majority of people who became Christians came through a "ran over the cat and killed it" experience. Things were going fine, and then a loved one got killed in a car accident, or they had cancer or financial collapse. It was some sort of crisis that came into their life that turned them to God.

Are all of these people weak-minded people? Do they need a crutch to go through life? Well, not really. They're ordinary people like you and me. They had a crisis that brought them to their knees and made them look upward.

I didn't come through that experience. My crisis was kind of different. At the age of 22, I had attained everything I wanted to attain—every material possession. I was incredibly happy. But I could see I was part of the ultimate statistic: ten out of ten die. I realized that everything I love was going to be ripped from my hands by death. So life took on a sense of futility. I thought to myself, why am I waiting around to die? It was like I had this big happiness bubble, and I was waiting for the sharp pin of reality to burst it. That was my crisis; and if you're not aware of *your* crisis today, let me give you one.[335]

Do you realize that you've sinned against God? Do you realize you've violated His Commandments? Some of you here today are here because you violated civil law. You have to face a judge. If there's a fine to be paid and you can't pay the fine, then you'll suffer the consequences.

You and I have violated another law—the Law of God, the Ten Commandments, so what I'm going to do is read the Law to you, and you can be the judge as to whether or not you're innocent or guilty. "You shall not commit adultery."[336] Ever done that? You say, "No. No, I've never done that." Folks, listen to what Jesus said. He said, "But I say to you, whoever looks upon a woman to lust after her has committed adultery already with her in his heart."[337] Ever done that? Ever looked with lust? You say, "Are you kidding? That's all I ever do. Guilty of that one."

"You shall not steal."[338] Ever stolen something? You say, "Yeah . . . just little things." Then you're a thief, irrespective of the value of that which you've stolen. "You shall not bear false witness."[339] Ever told a lie? You say, "Yeah. Fibs, white lies." Come on, have you ever borne false witness? I'm not talking about "discretion." I'm talking about telling a *boldfaced* lie. You say, "Yeah, I've done that. Guilty." Well, then, by your own admission you're a lying, thieving, adulterer at heart, and the Bible says, "All liars will have their part in the lake of fire."[340] No thief will inherit the kingdom of God, neither will adulterers.[341]

Have you ever used God's name in vain? Ever used His name as a cussword? Think about it. God gave you life. He gave you a brain to think with, eyes to see with, ears to hear with, taste buds to enjoy good food with. He lavished His goodness upon you, and you used His name as a cussword to express disgust. That's called "blasphemy." Blasphemers will not inherit the kingdom of God, and bear in mind there's been two witnesses to your crimes: Your conscience. Conscience means "with knowledge." *Con* is "with," *science* is "knowledge." Every time you violate God's Law, there is an inner witness—an impartial judge in the courtroom of your mind that says, "Guilty."

And the second witness to your crimes is Almighty God. The Bible says, "The eye of the Lord is in every place, beholding the evil and the good."[342] You think God can form the human eye, with 137 million light-sensitive cells (there's not a camera on the face of this earth that is anywhere near the sophistication of the human eye) . . . do you think God formed the eye, *and He's blind?* No, God sees everything, including your thought life and deeds you do in darkness,[343] and the Bible warns this righteous Judge of the Universe "will bring every work to judgment, including every secret thing whether it's good or whether it's evil."[344]

---

335 Remember that you are trying to alarm your hearers to their terrible danger. They live in a dream world of unreality. With God's help, you are trying to awaken them.
336 Exodus 20:14
337 Matthew 5:27,28
338 Exodus 20:15
339 Exodus 20:16
340 Revelation 21:8
341 See 1 Corinthians 6:9,10.
342 Proverbs 15:3
343 See Psalm 139:11,12.
344 Ecclesiastes 12:14

So think about this. If you die today, if your heart gave out right now and you stood before God, would you be innocent or guilty? You know you'd be guilty. You're like me . . . you're like the rest of us. Would you go to Heaven or Hell? The Bible makes it very clear: "All liars will have their part in the lake of fire."[345] You would be in big trouble if you died in the state you're in. Eternal damnation—where God cuts you off from life and light and laughter and pleasure, gives you "anguish," the Bible says. Read the Book of Romans, chapter two. He will give *anguish* to every soul that continues in sin.

So what should you do? The Bible says this Judge of the Universe is rich in mercy to all who call upon Him.[346] God in His goodness provided a way for you to be forgiven. He paid your fine in the life's blood of His Son. God became a person in Jesus Christ[347] and gave His perfect sinless life as a sacrifice for the sin of the world. The first time John the Baptist saw Jesus, he said, "Behold the lamb of God who takes away the sin of the world."[348] In the Old Testament the Jews had to bring a spotless lamb—a lamb without blemish—and shed its blood for a temporary covering for their sin. But God provided the Lamb of God, the sinless Lamb, a perfect human whose blood once and for all can wash away the transgressions of all humanity.[349] The Bible says, "For God so loved the world that He gave His only begotten Son, that whoever believes in Him should not perish but have everlasting life."[350] "Christ has once suffered for sins, the just for the unjust, that He might bring us to God."[351] The reason for His death and resurrection was so that your case could be dismissed on the Day of Judgment. It was so that death could lose its sting on your life and the grave could lose its victory,[352] and the moment you put your faith in Jesus Christ, the moment you repent, the Bible promises you'll pass from death to life.

Folks, please get right with God today. Call upon the name of the Lord. The Bible says that whoever calls upon the name of the Lord will not be disappointed. You won't be. You will pass from death to life. You have God's promise on it. You may not be here tomorrow. Every single year 42,000 Americans are killed in car accidents. Every single day, 1,500 Americans die of cancer. There's all these other ways you and I could die. Today could be your last day on the earth, so get right with God before you pass into eternity. Just cry, "God, forgive me. I'm a sinner," and willfully put your trust in Jesus Christ.

* * *

The Chickens and the Curve

A certain city years ago was having trouble with a portion of road outside the city limits. There was a curve on the road, and cars or vehicles would come along the road and go so fast they would flip, and people would be killed. It was happening so often the town's council met and had a meeting, particularly to find out what they could do for it.

---

345 Revelation 21:8

346 Ephesians 2:4

347 "And without controversy great is the mystery of godliness: God was manifest in the flesh, justified in the Spirit, seen of angels, preached unto the Gentiles, believed on in the world, received up into glory" (2 Timothy 3:16).

348 John 1:29

349 Hebrews 10:9–14

350 John 3:16

351 1 Peter 3:18

352 See 1 Corinthians 15:56.

What they decided to do was to put signs along the road warning people to slow down, which said "Dangerous...Slow Down!" Nobody slowed down. People kept going around that curve and getting killed.

So one of the council members had an interesting thought. He said, "Let's put a bunch (or a flock) of chickens on the side of the road, and let them be 'free-range' chickens, and see what happens."

So that's what they did. They put this bunch (or flock) of chickens on the side of the road, and drivers came screaming, saw the chickens, slowed down, went around the curve slowly, and the death rate went way down.

Isn't it interesting? These people were more concerned about the life of a chicken than they were about their own life.

Some of you are the same today. You're not at all concerned about your eternal salvation. You don't think about your death or where you'll spend eternity. You're more concerned about what you're going to have for dinner tonight than you are where you're going to spend eternity.

I heard a true story of a young guy. This kid befriended a chicken when it was really little, and fed it by hand. As the chicken grew, it became very close to this young boy. In fact each day he would go out after school, he'd take some wheat in his hand, hold it like that and this chicken would leave the other chickens, jump up on a fence and he'd feed it by hand. It happened every day.

One day, Christmas Day, the father, not knowing what his son had been doing with the chicken, came out to grab a chicken and kill it for dinner. Know what? As he stepped up to the chicken coup he found that he had a volunteer. One chicken left the other chickens, jumped up and he just grabbed it, chopped its head off and thought, *Man, that was great. No problem at all.*

That's what some of you are like when it comes to the issue of death—that's what you are doing by your inaction. You are thoughtless when it comes to your eternal salvation. You're just laying your head down, so to speak, and letting the chopper come upon you.

If you were being pushed toward a thousand-foot cliff, inch by inch, wouldn't you stop for a minute and say to the person pushing you, "Why are you pushing me? Is there anything I can do to stop you pushing me?" Wouldn't you do that? Of course you would! And yet day by day, you're getting closer and closer to that chopper coming down, to that cliff of eternity, and you're not at all concerned.

The day will come when death will arrest you, and you'll stand before a Law you've violated, and a Judge will condemn you. What I want you to do is listen carefully today. Just open your heart. Be patient with me as I read you that Law in the hope that you'll suddenly become concerned, and see that you better do what God says, if you want to preserve that life.

The Bible says, "What shall it profit a man if he gains the whole world and loses his own soul?"[353] So please listen carefully. The Bible tells us that you and I have violated the Law. It's an eternal Law. It's a Law you're very familiar with. It's called the Moral Law, or the Ten Commandments.

353 Matthew 16:26

It's written on your heart via your conscience—God has given light to every man.[354] There is something in you that says you shall not steal, you shall not kill, you shall not lie, you shall not commit adultery. Something in you says God should have some preeminence in your life. It's written on your heart. I know that to be true because before I was a Christian, I was a surfer, and I was following this car one day—on the back of the car it had a sticker and the sticker just said two words: "God First." And you know what? I couldn't wait to get past that car. I thought, " Eck...I don't want that stuff." It made me feel guilty, because I knew in my heart that God should be first in my life. Why? Because God has inscribed His Law upon my heart via the conscience.

Okay—Is God first in your life? Do you love the God who gave you life, who gave you breath, who gave you being? Do you love the God who gave you eyes to see with, ears to hear with—ears that can hear music...the creation of God? Do you love God? Do you love the God who gave you taste buds to enjoy the food He gave you? Because everything that God gave you in life came by the hand of God, the goodness of God, and the Bible says that none of us have kept the First of the Ten Commandments. Not one of us! The Scriptures say, "There's none that seek after God."[355] I'm as guilty as you. The Scriptures say, "We're all like sheep that have gone astray. We've turned everyone to his own way."[356]

How about some of those other Commandments: "You shall not bear false witness."[357] Have you ever lied? Have you ever told a lie? Then you're guilty of lying. You are a "liar" by the definition of the word. If you lied, you're a liar, and the Bible warns that "all liars will have their part in the lake of fire."[358]

"You shall not steal."[359] Have you ever stolen something, irrespective of its value? Then you're a thief, and the Bible says, "Thieves will not inherit the kingdom of God."[360]

"You shall not commit adultery." Maybe you haven't committed adultery, but listen to what Jesus said: "Whoever looks upon a woman to lust after her has committed adultery already with her in his heart."[361] Have you ever done that? Ever looked with lust? Of course you have. You're an average human being. You're normal. The Bible says we "drink in iniquity like water."[362] We love the darkness.[363] We hate the light.[364] We love to sin. That's what the Bible says. We're sin-loving creatures.

I remember I saw a card many years ago (some sort of humorous card). It said, "Everything I love is illegal, immoral, or fattening." Isn't it true? We love that which is immoral. If you don't believe it, look at television. Look at the movies. Go into your video store. Look at the type of magazines we have. We love uncleanness, and the Bible says, "God will bring every work to judgment including every secret thing whether it is good or evil."[365] There's certain communities in our society that say

---

354 See John 1:9.
355 See Romans 3:11.
356 Isaiah 53:6
357 Exodus 20:16
358 Revelation 21:8
359 Exodus 20:15
360 1 Corinthians 6:10
361 Matthew 5:27,28
362 Job 15:16
363 See John 3:19.
364 See John 3:20.
365 Ecclesiastes 12:14

you've got no right to judge what goes on in my bedroom, in the privacy of my own home. There's no such thing as "the privacy of your own home." There's no such thing as "the privacy of your own mind," because God sees your thought life. He sees what you think about. He sees what you do in darkness. Nothing is hidden from the eyes of Him to whom we have to give an account.[366] And you and I will stand before this God whose Law we've violated, and give an account of "every idle word" we've spoken.[367]

Folks, if you've got a tender conscience today, you will know that you've sinned against God, and you know that you desperately need His mercy. The Bible tells us that God is rich in mercy. He sent His Son, born of a virgin—Jesus Christ, a perfect sinless man, God in human form—to suffer and die for the sin of the world. If you've never understood it before, two thousand years ago when Jesus Christ was on the cross, what He was doing was taking the punishment for the crimes that you and I have committed against God. He was paying the fine in His life's blood, so you and I could leave the courtroom.

Jesus told a story. It's called a parable of The Prodigal Son. It's about a young man who went to his father and said, "Father, give me my inheritance. Give me the money owed to me." So his father gave it to him.

The Bible says that he went to a far country away from his dad. And he took that money and he spent it on "riotous living," and on prostitutes. He enjoyed "the pleasures of sin" but it was but for a season.[368] Only for a time, because the Scriptures say a famine came upon that land. His money ran out. His friends left him. The only job he could get was feeding pigs, and as he sat in that pigsty the Bible says he began to see that he was desiring to eat pig food—that filth that sat before him! He wanted to pick it up with all its filth and begin eating it.

The Scriptures say he came to his senses and said, "Even my father's servants have it better than I have." He said, "I'll go back to my father and say, 'Father I've sinned against heaven and in your sight, take me on as your hired servant.'"

So he got out of the pigsty. He came to his senses. He went back to his father and the Bible says, while he was yet afar off, his father was looking for him. His father ran to him, fell upon him, kissed him, put a robe around him, a ring on his finger, and put shoes on his feet. And he said, "My son was once dead. He's made alive." And they had a feast.

That's a picture of you and I. Our appetites are unclean. We desire that which is unclean. Folks, come to your senses and realize that the day will come that you will die. As I've said, that Law we went through will try you. Death will arrest you, the Law will try you, and the judge will condemn you, if you do not do what the Bible says and turn from your sins.

Now remember that parable. The father was looking for his son. That's what God is waiting for you to turn to do—to come toward Him. He will meet you halfway. He'll put a robe of righteousness upon you. He'll cleanse you of sin. He'll give you an inheritance. He will clothe you. The Bible says God Himself will rejoice that you were once dead and you've been made alive.

That's what happens when someone becomes a Christian. They pass from death into life. So think of that dumb chicken jumping up on that fence waiting for that chopper to come down. Don't be like that. Come to your senses. Realize, unless you do what God says, death will seize upon you.

---

366 Hebrews 4:13
367 See Matthew 12:36.
368 See Hebrews 11:25.

Every day in the United States, 1,500 people die of cancer. Every year just under a million die of heart disease. Every year 42,000 Americans are killed in car accidents. Forty-two thousand every year! The next car you get into may be your coffin. Every single day, 150,000 pass from time into eternity. Death seizes upon them.

So, folks, do what the Bible says: repent—turn from your sins and trust in Him who suffered and died on the cross for you, Him who rose again on the third day to defeat death. And the moment you do that, the moment you with an honest heart come before God and say, "God, forgive me. I'm a sinner. My heart's desires have been unclean. 'Create in me a clean heart O God, renew in me a right spirit'"[369]—God will do it. And you'll pass from death into life.

If you've got a Bible, folks, that's God's Word to you. Some people say the Bible's full of mistakes. Yes, it is. The first mistake was when man rejected God. Don't make that same mistake. Others say men wrote the Bible. Of course men wrote the Bible. It didn't evolve. When you write a letter, do you write the letter or does your pen? Well, you write the letter; the pen is the instrument you use. And God used men as instruments to pen His letter to humanity. All that whole Bible is, is a promise from God that He would destroy death. That's the Old Testament. The New Testament tells us how He did it—through the suffering death and resurrection of Jesus Christ.

Folks, for that Bible to make sense, you must be born of the Spirit. You must have God open your understanding. You know, most of you consider yourself reasonably intelligent people. Well, can you speak Chinese? It's a complicated language. Can you read or speak Chinese? I could bring a four-year-old Chinese child here and shame you, because the Chinese kid could read and speak fluent Chinese. Why? Because he's born into the family. Folks, when you're born into God's family the Bible will come alive. You'll speak the language of God. You'll understand His Scriptures.

So I want to thank you for listening to me today, and I'm going to come along the line in a moment and offer you a free CD called "What Hollywood Believes." It's completely free. Please feel free to take it. Thank you for listening.

* * *

The Fly

Have you ever noticed what a fly does when it lands on your plate after a meal? What's the first thing it does? Well, if you study it, it *cleans* itself. You watch its little hands go over its head. Then it'll clean its wings. Then it'll clean its legs. Clean machine is the fly. Now if you study man, watch what he does. He cleans his fingernails. He cleans his teeth, shampoos his hair, soaps his face; he cleans his clothes. Clean machine is man. But you follow him and see what his appetites are really for. Just like you follow a clean little fly and you see what it really desires. It's an unclean thing, the fly.

The Bible says, speaking of you and I, "We're all as an unclean thing and all our righteous deeds are as filthy rags in the sight of God."[370] So what does that mean, "We're all as an unclean thing?"

---

369 Psalm 51:10
370 Isaiah 64:6

Because I guarantee if I came along the line today and spoke to you and said, "Do you see yourself as an unclean thing in the sight of God?" you'd say, "What are you talking about? I'm a basically good person." The reason we feel like that is because we're like the religious people in the time of Christ. The Bible says this: "They went about to establish their own righteousness being ignorant of the righteousness which is of God."[371] What does *that* mean? Another word for "righteousness" is the word "goodness." These people went about to establish their own goodness being ignorant of the goodness which is of God. Well, let me put it another way. A little girl was once looking at a sheep as it ate grass, and she thought to herself, "That sheep looks nice and clean against the green grass." Then it began to snow and the girl said, "Boy, that sheep looks *dirty* against the white snow." It was the same sheep but there was a different background.

When you or I measure ourselves by the standard of man's goodness—by his righteousness—we come up clean. I mean, my life compared to Adolph Hitler's makes me seem a virtuous person. There are plenty of people worse than me, I can guarantee you that.

The Bible tells us on the Day of Judgment the backdrop will not be the standard of man, but the perfect white righteousness of a holy God. So what I'm going to do this morning is just very quickly go through some of God's Commandments to show you His standard of righteousness—what God expects of you and I.

The First of the Ten Commandments says, "I am the Lord your God, you shall have no other gods before Me."[372] That means God is your Creator, whether you believe in Him or not. He is your God. You have a God—the One who made you, who formed you in the womb, who gave you the personality you've got, gave you the color eyes you've got, the color hair, your height. Everything was given to you by God. We're not an accident.

He said, "You shall have no other gods before Me."[373] That means the One who gave you life commands that *He* be the One you love above all other things. You shall love the Lord your God with all of your heart, mind, soul, and strength, and love your neighbor as much as you love yourself.[374] Have you kept that Commandment? Have you obeyed it? Can you say, "I'm righteous as I stand before that Commandment?"

The Bible tells us no one can say that. It says there is none that seeks after God. The Bible tells us we're unthankful, we're unholy.[375] We don't care about God. We use His name as a cussword. We hold our fist up in rebellion of the God of Heaven and say, "God, it's my life. I'll live it as I want!" The Bible tells us our natural mind is a place of hostility when it comes to God and His moral government.[376] We don't want God telling us what to do. That's called "rebellion," and it's a transgression of the First of the Ten Commandments.

---

371 Romans 10:3
372 Exodus 20:3
373 Ibid
374 Matthew 22:37,38
375 See 2 Timothy 3:3.
376 See Romans 7:8.

The Second says, "You shall not create a graven image…you shall not make yourself a graven image."[377] That is, you should not make a god to suit yourself. You should not create a god that you feel comfortable with. In Bible times, they did that all the time. They'd shape a golden calf—they would make some sort of heinous idol. People still do it today. They make a physical idol and they bow down to worship this dumb idol that has no eyes, no ears, and no brain. It's an inanimate object. It's stupid. But we do the same thing in our minds. We don't create a god with our hands, but we create a god we feel comfortable with. We shape him to conform to our sins.

I did it before I was a Christian. I had a god I prayed to every night for ten years. I could rattle through The Lord's Prayer in about ten seconds. It meant nothing but a sleeping pill, a habit. I prayed to the god of my own creation—a non-existent god, a figment of my imagination, a god that had no sense of justice or righteousness or truth. That's called "idolatry," when you have a god to suit yourself. It's a violation of the Second of the Ten Commandments.

Let's skip over to the seventh: "You shall not commit adultery."[378] Now, maybe you've not committed adultery, but guys, listen to what Jesus said. He said, "But I say to you, whoever looks upon a woman to lust after her has committed adultery already with her in his heart."[379] You see, God requires truth in the inward parts.[380] The Bible says, "The Law is spiritual."[381] That means civil law is limited to what it can do when it comes to your crimes. It can't see your thought life. But the Bible tells us God goes right through to the thought life and He sees hatred as murder. Whoever hates his brother is a murderer.[382] He sees lust as adultery.

Nothing is hidden from the eyes of Him to whom we have to give an account.[383] Have you lied or stolen? Then you're a lying thief. Can you see how the Law shows us God's perfect standard of righteousness? That's the standard God's going to judge you with on the Day of Judgment.[384] So what will you do when all your secret sins come out?[385] When you realize in your heart the Bible's right when it says, "All liars will have their part in the lake of fire."[386] God is angry at sin. God is angry at sinners.

The Bible says that all nations that forget God, He'll turn into Hell.[387] "His wrath abides upon us"[388] because of our transgressions, because we've violated His Law. We're like a heinous criminal who slit the throat of a number of sweet young ladies after he raped them, and he's standing before a good judge. Is that judge going to have a benevolent attitude toward a criminal? No, he'll be *angry*. He'll bring down the gavel. He'll say, "I'm going to send you to the electric chair!" He's angered by the transgression of the criminal who stands in front of him. And the Bible says God's wrath "abides upon" each of us.

---

377 Exodus 20:4
378 Exodus 20:14
379 Matthew 5:27,28
380 Psalm 51:6
381 Romans 7:14
382 1 John 3:15
383 See Hebrews 4:13.
384 See James 2:12.
385 See Ecclesiastes 12:14.
386 Revelation 21:8
387 Psalm 8:17
388 John 3:36

But the wonderful thing is that this God who is just and holy is benevolent. The Bible says "He's rich in mercy"[389] and He "commended His love toward us in that while we were yet sinners, Christ died for us."[390] God became a human being in Jesus Christ and gave His life as a sacrifice for the sin of the world. He paid for our sins in His life's blood. He paid the fine. "For God so loved the world that He gave His only begotten Son that whoever believes in Him should not perish but have everlasting life."[391] Through the suffering death and the resurrection of Jesus Christ, God can remit your sins. He can dismiss your case. He can commute your death sentence and save you from everlasting damnation—eternal Hell, because of what Jesus Christ did on the cross.

Now the Bible says, "God commands all men everywhere to repent because He has appointed a day in which He will judge the world in righteousness."[392] Folks, do turn from your sins today. Don't wait until tomorrow. A hundred and fifty thousand people die every single day.[393] Today, you could die . . . I could die. It's a horrible thought, but it's the truth. We don't really believe that. But you can guarantee the 150,000 people who died yesterday didn't believe it either. So think soberly about your eternal salvation. Repent and put your trust in Jesus Christ. Don't just *believe* in Him; *trust* in Him like you trust a parachute.

The moment you do that you have God's promise, you'll pass from death into life. You'll pass from darkness into light. The Bible will come alive. Instead of God being far off, He'll be your closest friend. He says He promises to dwell within you through the power of the Holy Spirit. He'll seal you with His Holy Spirit and make you a new person. You'll be born again with a new heart and new desires.

* * *

The Persistent Spider

A number of years ago, we had a very persistent spider put a web across the front of our home. I'd go out there with a broom each morning and brush it down. Next morning the spider's spider web was there again. It happened three mornings in a row. After the third morning I thought to myself, *You know, I think my brain is bigger than the brain of a spider. I can outwit this beast.* So what I did is, I got my youngest son and I said, "You take this little stick here. Come with me, and we're going to deal with this problem." So we went there. I had a can of insect repellent in my hand. I said to my son, "You tap this web, just like a little vibration on the web while I very skillfully make the sound of a fly in distress." So that's what we did. He went *tap, tap,* and I went "*Zzzzzz.*" After about ten seconds of that, suddenly a spider comes out from its hiding place, yells to itself, "Lunch!," and we killed him with an insect repellent. We dealt with a problem.

Now, folks, if you look across the face of America, in fact the whole face of the world, you'll find a great web of corruption. We've got a problem with murder. We have rape. We have greed, anger, lust, pride, jealousy—all these things, all this corruption. What's the cause? When some guy rapes ten

389 Ephesians 2:4
390 Romans 5:8
391 John 3:16
392 Acts 17:30,31
393 Deaths per day worldwide: 153,558 (U.S. Census Bureau, 2004)

women and murders them, or some kid takes a gun and shoots six of his friends at school, in come the psychologists and they say, "What is the profile of a person like this? What causes somebody to go astray like this, and do something so horrific?" Well, the Bible doesn't say there's any profile. It puts its finger on the problem. The Bible says, "It's the spider of the sinful, wicked human heart."

But you know something? That spider hides. He's got a hiding place. I know if I came across this line this morning and walked down the line and said, "Do you agree with the Bible when it says your heart, your nature is wicked?" You'd probably say, "Oh, no. No, no. I'm basically a good person." So what I'm going to do with your help, your cooperation hopefully, is take the stick of God's Law—those Ten Commandments—while I'm making a buzzing noise of the gospel, and I'm going to see if we can make that spider of that sinful, wicked human heart reveal itself, so that we can kill it with a

fly spray of the gospel. So please listen carefully.

Have you ever told a lie? Or have you ever stolen something? Have you violated that Commandment, "You shall not bear false witness,"[394] or "You shall not steal"?[395] Have you ever done that? You say, "Ah, yeah. I've taken a few little things in my life. Yes, I have told a few white lies. I have lied." Well, if that's true, you're a lying thief. You've violated those Commandments—the Ninth and the Eighth. Have you ever desired something that belongs to somebody else? Well, that's called "covetousness." It's a violation of the Tenth Commandment, "You shall not covet."[396] It means to have greed in your heart. What about the Seventh Commandment: "You shall not commit adultery"?[397] You say, "I've never done that." But listen to what Jesus said. He said, "You've heard it said by them of old you shall not commit adultery, but I say to you whoever looks upon a woman to lust after her has committed adultery already with her in his heart."[398] Have you ever done that? Well, then, you've violated the Seventh Commandment. God sees you as an adulterer.

Sixth Commandment: "You shall not kill."[399] Now maybe you haven't killed anybody. But Jesus said if you get angry without cause you're in danger of judgment.[400] In fact the Bible says, "Whoever hates his brother is a murderer."[401] So if you've ever hated someone, you've committed murder in your heart. God sees you as a murderer.

Or have you kept the First and Second of the Ten Commandments: "You shall have no other gods before Me"?[402] Do you love the God who gave you life? Or have you used His name as a cussword to express disgust—taken His name in vain? That's called "blasphemy." Or perhaps you've done what I did before I was a Christian. I made a god in my own image. I created a god to suit myself and

394 Exodus 20:16
395 Exodus 20:15
396 Exodus 20:17
397 Exodus 20:14
398 Matthew 5:27,28
399 Exodus 20:13
400 Matthew 5:22
401 1 John 3:15
402 Exodus 20:3

violated the Second of the Ten Commandments. I prayed every night for ten years as a non-Christian. I'd say The Lord's Prayer—rattle it off as a sleeping pill. But it wasn't to the God of creation; it wasn't to the God of the Bible. It was to the god of my own imagination, a god who didn't exist. A god who had no sense of right and wrong, a god who wasn't going to punish people on Judgment Day, who didn't create Hell—a non-existent god. That's called "idolatry."

Have you kept the Sabbath holy? Have you honored your father and mother? As you look at those Commandments you say, "Man, I've blown it." From God's point of view, if you've broken those Commandments, you're an ungrateful, unthankful, blasphemous, lying, thieving, murderous, adulterer at heart. And the Bible's testimony is true: our hearts are deceitfully wicked.

Can you see how tapping your heart with those Commandments makes the spider of the true self reveal itself? And there's a reason it needs to be done. It's so that sinful nature can be killed. That's what the gospel does. We become crucified with Christ when we become Christians. And when a nation turns to God, it gets rid of the web of corruption.

If you study revivals in history, you'll see in the Welsh Revival when the Spirit of God transformed people and they were all made new people in Christ—they were born again—the police had no work to do. Seriously, they were bored. They didn't know what to do because there was no corruption. They even had to retrain the miner's donkeys. The miner's donkeys were used to hearing verbal abuse and cussing. These men became Christians and became kind and loving, and their mouths became clean. And so the donkeys had no clue what to do. They had to retrain them as to what to do.

But the reason you should get right with God today isn't to clean up the corruption in our nation. The primary reason is that God has "appointed a day in which He'll judge the world in righteousness."[403] And if you stand before Him as a liar or a thief, you're in big trouble. If you stand before Him as an adulterer or murderer or a blasphemer, you're in big trouble because the Bible says, "All liars will have their part in the lake of fire."[404]

You cannot enter heaven if you have sin in your heart. Nothing defiled will enter heaven. Now, you don't want to go to Hell. I don't want you to go to Hell. God doesn't want you to go to Hell. The Bible says "the god of this world,"[405] Satan, wants to "kill, steal, and destroy."[406] He will take you to Hell if you serve him, and you're serving him if you're serving sin. The Bible makes it very clear.[407]

So what are you going to do on the Day of Judgment when you stand before God and "give an account of every idle word"?[408] The Bible warns, "It's a fearful thing to fall into the hands of the living God."[409] As I said, God doesn't want you to end up in Hell.[410] Do you know what He did for you? Almighty God, the Creator of the universe, made a way for you to be forgiven. He sent His Son, born of a virgin. Jesus Christ was God in human form, and this perfect man gave His life as a sacrifice for the sin of the world. You know that Christ died for our sins, but when He was on that cross,

---

403 Acts 17:31
404 Revelation 21:8
405 See 2 Corinthians 4:4.
406 See John 10:10.
407 See John 8:44; Acts 26:18; 2 Timothy 2:26; 1 Peter 5:8.
408 See Matthew 12:36.
409 Hebrews 10:31
410 See 2 Peter 3:9.

the Bible tells us He was actually *purchasing* redemption for us. He was purchasing God's forgiveness for us. He was paying the fine in His life's blood, so we could leave the courtroom.

Now some of you are going to stand before a judge today. Some of you are going to stand before a law you've violated, and you're going to hear that you have to pay a fine. Imagine being guilty and someone you don't even know coming in and paying that fine for you. Wouldn't that be wonderful? That means you're free from the law. The judge says, "Fine's paid. You're out of here."

Well, that's what God did for you in Jesus Christ. Jesus suffered for sins once, "the just for the unjust, that He might bring us to God."[411] "For God so loved the world that He gave His only begotten Son that whoever believes in Him should not perish but have everlasting life."[412] Through the suffering death of Jesus Christ, through His resurrection (He rose from the dead on the third day), God can now forgive every sin you've ever committed, and grant you His gift of everlasting life.

Folks, please come to your senses today. Let God destroy that old nature through the power of the gospel. Let Him transform you. Let Him bring you out of darkness into light. Allow Him to grant you the gift of everlasting life. Because that's what He offers you today. To receive it, you must repent—that is, turn from all sin. It means crying out, "God, forgive me. I'm a sinner. Create a clean heart in me, oh God, renew within me a right spirit."[413] And, folks, if you'll come before Him and do that, and place your trust in Jesus Christ, God will forgive every single wicked thing you've done. All those deeds you did in darkness, all those unclean thoughts...God can wash them away and remove your sins "as far as the east is from the west."[414] Please do it today. Please call upon His name today because you may not have tomorrow.

Who knows what could happen in our city, and our state, and our nation? Who knows how long your heart is going to stay beating? You could be whisked into eternity in a heartbeat, if God lost patience with you. So "today if you hear His voice, don't harden your heart."[415] Call upon the name of the Lord and the Bible promises, "Whoever calls upon the name of the Lord shall be saved."[416]

* * *

The World's Most Valuable Commodity

Some years ago I turned on the television and saw the newsreader say, "Today we're going to talk about the world's most valuable commodity." I thought, *What is the world's most valuable commodity? What is it? Is it gold? Is it oil? Is it diamonds?* And then he said something interesting. He said, "It's information." I thought, *That guy is right. Information is the world's most valuable commodity.* If you know where the oil is in the soil, or the gold, or the diamonds, you can sell that information. Information *is* the world's most valuable commodity.

---

411 1 Peter 3:18
412 John 3:16
413 Psalm 51:10
414 Psalm 103:12
415 Hebrews 3:7,8
416 Acts 2:21

Information can even save your life. If you're in a high-rise hotel and you smell smoke and you look on the back of the door, it'll tell you how to get out. It may say something like: Get low to the floor. If you breathe in, standing up you'll breathe in toxins and they will kill you with just one breath. And they say, "Don't turn to the left—that's a deathtrap. Turn to the right. Go to the end. Find an exit." Information can save your life.

The Bible is a book of information. An acronym for "Bible" is: Basic Instructions Before Leaving Earth. I know if I came down this line and asked you, "How does a man enter Heaven, how does a man or woman enter Heaven?" most of you would have the wrong information. You'd probably say something like this: "Well, you just have to believe in God and try to live a good life." Uh, uh. Wrong information. Dead end. The Bible says, "There is a way that *seems* right to man but the end of that way is death."[417]

This morning I'd like to give you some information. I'd like you to open your heart. You've got nothing to lose. It costs you nothing. All you have to do is set aside prejudice, and say, "Okay, what *is* the Bible's explanation on how to get everlasting life?"

What I want to do is to show you your danger. You see, if you're in a hotel you'll never read those instructions behind the door if you think things are fine. But if you smell smoke, you'll say, "This place is on fire!" You'll run to those instructions and they'll tell you what to do and what not to do. The Bible tells us that each of us is in great danger. Every single one of us is in great danger because we've sinned against God. We've violated His Commandments.

So let me go through just a few Commandments and see how you're going to do on the Day of Judgment. "You shall not bear false witness."[418] Have you ever told a lie? Have you broken that Commandment? Have you borne false witness? Have you lied? You say, "Yeah, I've told one or two." Well, you may not think it's serious, but God does. The Bible says lying is an abomination to the Lord.[419] That means it's incredibly distasteful to God to be a false witness. In fact the Bible says, "All liars will have their part in the lake of fire."[420] That shows how serious sin is in God's sight.

"You shall not steal."[421] Have you ever stolen something, irrespective of its value? Then you're a thief. Now listen to the voice of your conscience as we go through these Commandments.[422] I don't know about you, but if you had the experience of hearing a song from years ago, and suddenly a memory comes back—something you thought you'd completely forgotten; or a smell you smell, and suddenly *years ago*, this memory comes flooding into your mind—that's because your eyes, your ears

---

417 Proverbs 12:14

418 Exodus 20:16

419 Proverbs 12:22

420 Revelation 21:8

421 Exodus 20:15

422 Make continual reference to the conscience. It is the inner light that God has given to every man. It bears witness with the "work of the law" (Romans 2:15). The "work" of the Law is to bring the knowledge of sin and show sin to be exceedingly sinful (see Romans 7:13). As the Law does its work, the conscience is like a good judge, who, when listening to undeniable evidence of guilt, sits in affirmation of retribution.

have recorded everything you've seen, heard, and done . . . and your memory banks. You just don't recall them until that sound, that music brings them to recollection.

The Bible tells us we haven't done anything that God hasn't seen. Everything is going to be brought to judgment. Jesus said, "There is nothing hidden that shall not be revealed."[423] All that sin you've committed will come out as evidence of your guilt on the Day of Judgment. Now listen to this one: Jesus said, "You've heard it said by them of old, you shall not commit adultery, but I say to you, whoever looks upon a woman to lust after her has committed adultery already with her in his heart."[424] How's your conscience there? What does it say to you? It's an impartial judge in the courtroom of your mind. If you're an average red-blooded guy it should be saying, "Guilty. I've broken that Commandment. I'm guilty of adultery in the sight of God," because when we sin we sin against God.

Often we justify ourselves by saying, "But it doesn't hurt anybody." Of course pornography and lust doesn't hurt anybody. No one gets hurt because you look at a photo. But when you lust, you commit adultery in God's sight and sin against Him, and you provoke His anger.[425]

The Bible also says, "You shall not kill."[426] But did you know the Scriptures say whoever hates his brother is a murderer? Have you honored your parents? Have you kept the Sabbath holy? Have you ever used God's name in vain? Used the name of the God who gave you life as a cussword? Have you been greedy or covetous—desired things that belong to other people? Have you put God first? Or have you created a god in your own image, a god you feel comfortable with?

When we look at those Commandments . . . man, we're guilty! Every single one of us, and the Bible says when on Judgment Day all those sins come out as evidence of our guilt, we'll be condemned and end up in Hell. You've got God's promise of that. He will have His Day of Justice. So what are you going to do on that Day? Can you smell the smoke? Can you see your danger? Let me take you to the back of that door, so to speak . . . and look at what God says to do to be saved from that danger.

The Bible tells us, this same Judge of the Universe—the same vengeful,[427] angry, righteous, and holy God—is rich in mercy, and He sent His Son, born of a virgin. Almighty God became a human being and He gave His perfect life as a sacrifice for the sin of the world.

In other words, God became a person to suffer and die in your place and my place, to pay the fine, so we could leave the courtroom. "For God so loved the world that He gave His only begotten Son that whoever believes in Him should not perish but have everlasting life."[428] Folks, that's what God offers you through the suffering death and the resurrection of Jesus Christ. God loved you so much, He took the punishment for your sin upon Himself so you could be free from His anger. He paid your fine in His life's blood. Such is His love for you, and such is His love for justice. He fulfilled the demands of Eternal Justice through the suffering death and resurrection of Jesus Christ.[429]

---

423 Mark 4:22

424 Matthew 5:27,28

425 Romans 2:5; John 3:36

426 Exodus 20:13

427 See Psalm 94:1.

428 John 3:16

429 Labor the cross until you feel sure that your hearers understand what happened at Calvary. Millions know that Jesus Christ was crucified. They even know that He suffered and died "for the sin of the world." But they don't *personalize* it to themselves, and until they do that, they will never seek the Savior with a repentant heart. You are giving *knowledge* of the disease, so that the patient will see his fatal condition, and seek the cure. So you must link *knowledge* of their personal transgressions with the cure of the cross.

So there is provision for your forgiveness. The Bible says, "How shall we escape if we neglect so great a salvation?"[430] So what should you do when you smell that smoke, when you see your danger? Open the door of your heart. Get low. Get to your knees! "Humble yourself before Almighty God."[431] And don't breathe in that toxin of sin, because it will kill you. It'll take you to Hell. "God resists the proud and gives grace to the humble."[432] So you must walk with the lowliness of heart before Almighty God, and turn to the right. Turn to the path of righteousness. The Bible says, "In righteousness is the way of life."[433] God will "lead you in a path of righteousness."[434] And that path, that "way" is Jesus Christ. He said, "I am the way, the truth and the life; no man comes to the Father but by Me."[435]

So humble yourself, repent of your sins. That means confess and forsake them. Put your trust in Jesus Christ. Trust Him like you trust a parachute. Trust Him with your eternal salvation. Cry out, "God, forgive me. I'm a sinner. You're my Creator. You gave me life itself. I yield this life that I'm trying to keep from You, to You, to be my Lord and Savior. You're going to be my Boss from now on. I'll obey Your Word and love You." That's how you show that you love God—by *obeying* Him, and the Bible is God's revelation to you. It's "a lamp to our feet and a light to our path."[436]

If you want to know what God thinks, what He wants of you, read the Scriptures. Read the Bible. Folks, if you obey God, He promises to reveal Himself to you. Listen to this wonderful promise of God: Jesus said, "He that has My commandments and keeps them, he it is that loves Me. And he that loves Me will be loved of My Father. I too will love him and will reveal Myself to him."[437] So there you have it. "A way of life and a way of death."[438] If you want to stay in your complacency, it'll take you to Hell. But if you see your danger, follow that information that God's given us in His Word. Humble yourself and trust in Jesus Christ. Walk on a path of righteousness. Folks, the Bible says you'll be saved: "Whoever calls upon the name of the Lord shall be saved."[439]

* * *

The Wrong Apartment

Good morning. I want to have your attention for a few moments. My name is Ray, and I'm going to speak to you. I'd like you to know that I'm not a part of the court system, although I've been doing this for many years and the sheriff has assured me personally that I've a right to be here under my First Amendment freedom of speech rights. So I'd be grateful for your patience and your attention.

---

430 Hebrews 2:3
431 1 Peter 5:6
432 James 4:6
433 Proverbs 12:28
434 Psalm 23:3
435 John 14:6
436 Psalm 119:105
437 John 14:21
438 Deuteronomy 30:19
439 Acts 2:21

A number of years ago I was asked to speak in a city about 200 miles from my original city—the home I lived in. It was a Wednesday night and I went to speak to a group of people. I came back to the group apartments about two hours later, and noticed the people I was staying with had left a light on for me. So I opened the door, locked the three locks on the inside, went down the hallway into the bathroom, and looked around the bathroom and thought, *Man, these people are amazing. What*

they've done is renovated their bathroom in two hours! And then I realized I was in the wrong apartment. So I snuck down the hallway (I could hear people talking in the living room), unlocked the door, and snuck out.

I was sincere in my belief that I was in the right apartment, but I was sincerely wrong. There are people who in the name of tolerance say, "It doesn't matter what you believe, as long as you are sincere. That's all that matters; it doesn't matter whether you're a Hindu, Buddhist, Muslim, or Christian—as long as you are sincere."

Now, if you study Christianity you'll find that it is incredibly intolerant. It says it's not a matter of sincerity; it says sincerity must be coupled with truth. And it says this: There is only one way to God. And it's not through Islam; it's not through Hinduism, not through Buddhism. It's only through Jesus Christ. And there's a reason Christianity is intolerant, and that's what I want to share with you today. I trust you've got an open heart.

There is a reason that Christianity is intolerant of other religions. There is a reason why it says it's the only way to God, and this is the reason: The Bible says, "God has appointed a day in which He'll judge the world in righteousness."[440] That means God has a righteous standard—a perfect standard with which He'll judge humanity. He's a God of justice, a God of truth, and when He sees injustice, He must make it right. When a man commits murder, the man must be punished; a man rapes a woman, the man must be punished.

But God is so thorough, He's so righteous, so just and so good, that the Bible says He's not only going to punish murderers, He's going to punish those who desire to murder and never had opportunity. The Bible says, "Whoever hates his brother is a murderer."[441] In fact, the Scriptures tell us that if you get angry without cause you're in danger of judgment.[442] Now if you think that's a high standard, listen to this one. The Bible says, "You shall not commit adultery."[443] But listen to what Jesus said. He opened up the spiritual nature of the Law—what the intent of the Law is, that Moral Law. He said, "You've heard it said by them of old you shall not commit adultery, but I say to you whoever looks upon a woman to lust after her has committed adultery already with her in his heart."[444] If you've as much as lusted, the Bible says you've committed adultery. Now how are you going to do on

---

440 Acts 17:31
441 1 John 3:15
442 Matthew 5:22
443 Exodus 20:14
444 Matthew 5:27,28

the Day of Judgment, because the Bible says adulterers will end up in Hell? That's the standard God's going to judge us with. Or how about this one: "You shall not bear false witness."[445] Have you ever told a lie? If you have, the Bible says, "All liars will have their part in the lake of fire."[446] What will you do on that Day when God manifests all the secret sins of your heart? Or, "You shall not steal."[447] Have you ever stolen something? You say, "Yeah, just a little thing." Then you're a thief. You cannot inherit the kingdom of God. Ever used God's name in vain? You've taken the name of your Creator—the One who gave you life—and used His holy name as a cussword to express disgust. The Bible says, "The Lord will not hold him guiltless who takes His name in vain."[448] In fact, Jesus said, "Every idle word a man speaks, he'll give an account thereof on the Day of Judgment."[449] So what's going to happen to you when you stand before God and all those secret sins of your heart are revealed? All that lust, that selfishness, that gossip, those idle words that you spoke... the fact that God gave you life and you didn't give Him two minutes serious thought.

The Bible says, "You shall have no other gods before Me."[450] That means God should be the focal point of your affections. You and I are commanded to love God "with heart, mind, soul, and strength."[451] We're guilty before God. The Bible says when we stand before Him, if He gives us justice each of us would end up in Hell. So we're in big trouble—all of us. We're up the river Niagara without a paddle. We can't save ourselves.

So, can we go to Hinduism to save us? Hinduism cannot forgive sins. Neither can Islam. You talk to any Muslim and they'll say, "Oh, no, we don't have hope of eternal life. All we can do is say, 'Oh, God, I'm sorry for my sins,' but we don't have assurance that God forgives us. We have to wait until that Day." Now that's like saying to a judge, "Judge, I'm guilty of terrible crimes—rape and multiple murder—but, Judge, I want to tell you I'm really sorry." He's not going to let you go because you're sorry. He's bound by the demands of justice. And it's the same with God. Telling God you're sorry for your sins will not help you. No religion can help you.

There's only One who can help you, and that's Almighty God Himself. He provided a way for you and I to be forgiven. God became a person in Jesus Christ, and suffered and died on the cross for our sins. The Bible says of Jesus, [Heckler: Did you say that no religion can help you?][452] "He was bruised for our iniquities."[453] This was God in human form making a way for you and I to be forgiven. He was paying our fine in His life's blood. [Heckler: Did you say that no religion can help you?] That's why Christianity is unique. That's why it's exclusive because no religion provides a Savior.

Jesus said this: [Heckler: No religion, no religion, no religion, right?] He said, "The Son of Man has power on earth to forgive sins."[454] Folks, listen to this: the Bible says, "Neither is there salvation

---

445 Exodus 20:16
446 Revelation 21:8
447 Exodus 20:15
448 Exodus 20:7
449 Matthew 12:36
450 Exodus 20:3
451 Matthew 22:37
452 It was unusual to get a heckler at the courts (perhaps one every two months). In open-air preaching, hecklers can be a Godsend when it comes to helping to draw a crowd. However, they are not good if they heckle when the cross is being preached. This is the work of the adversary to try to bring confusion. If this happens, let him have his say, and then go back and labor the truth of the cross.
453 Isaiah 53:5
454 Matthew 9:6

under any other. There is no other name under heaven given among men whereby we must be saved."[455] [Heckler: What does the Koran say?] "Whoever transgresses and abides not in the doctrine of Christ, has not God."[456] Jesus Himself said, "I am the way, the truth and the life; no man comes to the Father but by Me."[457]

So realize your state before God. You and I are guilty criminals. Saying we are sorry can't help us, because we've violated the Laws of eternal justice. The only thing that can help us is the mercy of God, and that's been extended to us by God Himself through the person of Jesus Christ. So today, do what the Bible says: Repent and trust in Jesus Christ. Don't become a self-righteous religious person.

Now you know what self-righteousness is, and most religions are built on the doctrine of self-righteousness. That means they know they're guilty. A Hindu, a Muslim, a Buddhist, knows they're guilty. Their conscience says, "Lust is burning in your heart." [Heckler: Quit talking about other religions!] You've violated those Commandments, so what they try to do [Heckler: Quit assaulting other religions, you don't know... (Unintelligible).] is they try to get rid of their own sins. [Heckler: Quit assaulting other religions. They're all religions.] And they do it by good works and by suffering. Let me explain what I'm trying to say. Let me try to explain what I'm trying to say. [Heckler: You just got an opinion. Quit talking about other people, other religions like this.]

My name's Ray; what's your name? [Heckler: My name is Sean.] Sean? [Heckler: Yeah.] Sean, imagine this. If I said to you, "Sean, [Heckler: Yeah.] despite the fact that you're trying to stop me, I'm going to pay your fine today. Six hundred and fifty dollars, is it? [Heckler: I'll say, "I love you, Ray. Come on over." (Unintelligible.)] Would you like that? [Heckler: Absolutely.] Would you be grateful? [Heckler: Huh?] Would you be grateful? [Heckler: I would just be like, "Pay my fine."] Would you be grateful? [Heckler: Yeah, I think I would.] You would, wouldn't you? Because you don't like me and I'm still paying your fine. Folks, [Heckler: I'm not saying I don't like you. I'm just saying you keep on insulting other religions—that you preach your religion, and you just told us no religion can't save us.] That's true! [Heckler: No religion, including your religion.]

But what I want to try to say is this, Sean, is that God... [Heckler: We can only save ourselves, and that's what God wants us to do.] Yeah, you want to try to pay your own fine, go ahead, but the fine for sin [Heckler: Oh, I'll pay.] is everlasting damnation. That's the payment you've got to make. [Heckler: I don't live underneath that guilt. That's people like you, live underneath that guilt. I don't. Now my god...he wants me...(Unintelligible.)] Sean, I want to make a very interesting point about your fines. So just listen for a moment if you would, okay? Be tolerant, would you? [Heckler: But you've been talking for like thirty minutes. I like to talk too.] Give me a break! It's been nine minutes! Feel like thirty minutes to you? [Heckler: Okay you've been talking nine minutes and I'm like, "Equality, equality,"...just want to talk. I'm just, like, self-centered. I just want to talk as much as you.] I'll hold your place. You come here and talk. [Heckler: No. I'm talking from here.] Oh, you a bit nervous? [Heckler: Absolutely not!] Come out and talk! [Heckler: If I was nervous, do you think I'd be talking to you like this?] Yes! [Heckler: I'm absolutely not nervous.] You're not?

[Heckler: You're preaching Christianity. I'm saying, quit insulting all the other religions who have hundreds of years on Christianity. So quit your talking. Quit telling us to choose one over the other. Why don't we just believe in God? Why do we have to go through Jesus Christ? I think we can

---

455 Acts 4:12
456 2 John 1:9
457 John 14:6

transcend Jesus Christ and go straight to God. That's what god tells me. God tells me that he didn't send a human. He wants to talk straight to everybody. Christianity is like, "Pick—choose our flavor; pick—choose Baskin Robbins over Ben & Jerry's." No, no, no, no. Don't talk that over me if you don't know about him that much. That's an insult; that makes you look like you're not accepting, a phony, because a man of wisdom, a wise man would realize he doesn't know everything about every religion. So how can he say anything accusing other religions and say other people are damnation, are going to go to Hell because they don't believe in Christianity? That's ridiculous! That's ridiculous to say that if we don't believe what you believe that we're all going to go to Hell. No! God doesn't work like that. God did not put his message into human's hands and let him spread their flavor; no it's wireless, you go straight to God to us—period. We don't need you at the courthouse telling us we're liars and that we're all that kind of crazy stuff. Come on, everybody knows where we're at. Everybody knows where we're at. You got your issues, but I bet if we opened your book it's (horn sounded—unintelligible). Now you know. Now you're out here preaching to us. But that's what I'm saying. Don't sit there and condemn other people for their ignorance. And don't sit here and insult other people. Um, religions who have had hundreds of thousands of years of study just like yours has. You've probably only been studying your religion for about thirty years maybe at the most. There's people who have been studying religions for hundreds of years and they have just as much doctrine, just as much books as your religion. So don't sit there and say that everyone else (Unintelligible); don't sit there and convince me that I'm going to Hell because I don't believe what you do, all right? (Unintelligible. Some silence.) I'm spiritual and I know exactly what god tells me. I know one thing—that Jesus Christ is breaking down churches with music, all right? One more thing: the church is one that killed Jesus Christ, and I know one more thing that—Jesus Christ was brought up as this endorsement voice so everyone can sit there and say through Him is the only way of salvation. Through our faith in this moment, in our time, not that way (Unintelligible). You know what sense this makes? You know what I'm saying? You don't, do you? That's right, you got thirty years of that one way, of that one way of thinking. You're not talking. You have got to listen to everybody who talks different languages, who talks different (unintelligible). In some countries there ain't no such thing as Christianity. (Horn. Unintelligible.) War in Iraq, in the name of Christianity.]

Thank you, sir. Are you done? [Heckler: Absolutely.] Sean is guilty. He admits he's guilty. He's violated the law. Imagine if I paid his fine. Wouldn't that be a good thing for me to do for him? Well, that's what God's done for you and I. He paid our fine through the suffering death and the resurrection of Jesus Christ. He paid the fine for humanity in His life's blood. Now, if you don't want to partake in God's mercy, then you have to pay the fine yourself, and the Bible says there will be Hell to pay. If you've violated those Commandments, and you say. "I want to pay my own fine," the Bible says, "All liars will have their part in the lake of fire."[458] No thief, no adulterer, or fornicator will inherit the kingdom of God. And remember, Jesus said if you look with lust you've committed adultery in your heart.

But God is rich in mercy. He sent His Son to suffer and die on the cross. "For God so loved the world that He gave His only begotten Son that whoever believes in Him should not perish but have everlasting life."[459] So, folks, think seriously about your eternal salvation. Repent and put your faith in Him who died for you and rose again on the third day, and you'll pass from death into life. Thank

458 Revelation 21:8
459 John 3:16

you so much for listening. In a moment we're going to come down the line and offer you a free CD, a CD called "What Hollywood Believes."[460] Please feel free to take it. It will help you, and you'll find it fascinating.

* * *

The Wrong Key

A group of men were once in a warehouse playing poker. As they were playing (it was a very good game), one of the men said, "I can smell smoke!" And he looked behind him and there was smoke coming out of the door of the warehouse. His friend said, "Yeah, yeah . . . the place is on fire, but let's just finish this hand." He said, "No! Let's get out of here!" His friend said, "Let's just finish this hand. I got a key to get out. It's in my pocket. We can get out the back door." The guy says, "Okay . . . we can finish this hand. Be quick about it."

Precious seconds passed, and flames began to lick under the door, come up at the window. The guy said, "Hurry up, hurry up!" He said, "Okay. Look the key's in my pocket. Here it is on the table." He put the key on the table.

He said, "Let's finish this hand." More precious seconds went. Flames began coming in the win-

dow. Suddenly the man says, "Let's get out of here!" He grabbed the key, rushed to the back door, put the key in and turned around and said, "Ahhh . . . it's the wrong key!"

Now, folks, some of you really enjoy a good gamble, and your gamble is this: You're saying to yourself, "You know, if there's a God, if there's a Heaven, if there's a Hell, I've got the key to get in." You're totally confident you know the way to get out of your problem, and to get into Heaven. This is your key. You say to yourself, "As long as you live a good life, and you believe in God and you don't harm any other people, that's the key to get into Heaven." Uh, oh! *Wrong key!* That will leave you in a deathtrap. The Bible says, "There is a way that seems right to man, but the end of that way is death."[461]

Now I'm going to give you something that will provoke your thought today, that will help you smell the smoke—that will give you a sense of urgency to make sure you've got the right key.

I don't know if you realized it, but you've got a destiny with death. The reason death will come to you is because you've sinned against God. You've violated the Law and you'll stand before God. Death will arrest you, the Law will condemn you, and the Judge—God—will sentence you. I'm going to read you the Law so you can see the standard with which God will judge you.

---

460 Sean took one.
461 Proverbs 14:12

Most of you would consider yourself to be a good person. The Bible says that. It says, "Most every man will proclaim his own goodness."[462] We think we're good because we make a basic mistake. And that's this: we measure our self by a standard of man rather than by the standard of God.[463]

This is the standard of God—the Ten Commandments. I'm sure you're familiar with them. They're written upon your heart. "You shall not steal."[464] You know that's wrong; you've got a conscience. Let me ask you this question: Have you ever stolen something? It doesn't matter what it is, have you stolen something in your life? Then you're a thief. What about, "You shall not bear false witness"?[465] Have you ever told a lie? Then you're a liar, and the Bible says, "All liars will have their part in the lake of fire."[466] That's the standard God's going to judge with.[467] Or, "You shall not commit adultery."[468] Listen to what Jesus said. He said, "But I say to you whoever looks upon a woman to lust after her has committed adultery already with her in his heart."[469] Have you ever done that? What about the Tenth: "You shall not covet"?[470] Have you ever desired something that belonged to somebody else—husband, wife, boyfriend, girlfriend, stereo, whatever? Then you've violated the Tenth Commandment.

Have you ever used God's name in vain—you used God's name as a cussword to express disgust? That's called blasphemy, and the Bible warns, "The Lord will not hold him guiltless who takes His name in vain."[471] Folks, what are you going to do on the Day of Judgment, when all your secret sins come out? We're just talking about sins that are obvious. What about those deeds you've done in darkness, that nobody knows about? Well, the Bible says God sees darkness as pure light. He sees everything you think. He hears everything you say. Nothing is hidden from the eyes of Him to whom we have to give an account.[472] What will you do on the Day of Judgment? The Bible says you'll be condemned by that Law, and if God carries out your sentence, you'll end up in Hell for eternity. Well, that's not God's will. The Bible says, "He's not willing that any should perish."[473] It's not your will—something in you says, "Oh, I don't want to die!" So what will you do? What can you do? Well, you can enter a plea. "Okay, what will I plead? Innocent!" Oh, you dare not do that. It's obvious you're guilty; you know you're guilty. And if you plead innocent to a judge when all the evidence is there, he'll throw the book at you.

Exactly the same applies with God. You dare not plead innocent. You are guilty. You know you are guilty. Your conscience says you're guilty. The Law condemns you. Only thing you can do is plead "guilty," and throw yourself upon the mercy of the Judge. And the Bible says, "God is rich in mercy to all that call upon Him."[474] And the reason God can extend mercy to you is because He sent His Son,

462 Proverbs 20:6
463 Romans 10:3
464 Exodus 20:15
465 Exodus 20:16
466 Revelation 21:8
467 See Romans 2:12.
468 Exodus 20:14
469 Matthew 5:27,28
470 Exodus 20:17
471 Exodus 20:7
472 Hebrews 4:13
473 2 Peter 3:9
474 Ephesians 2:4

born of a virgin. Jesus of Nazareth was different than you and I. He was morally perfect. The Bible tells us this perfect man gave His life as a sacrifice for the sin of the world.

Jesus of Nazareth was crucified for the sins that you and I committed. We did the crime; He paid our fine in His life's blood, so we could leave the courtroom. "For God so loved the world that He gave His only begotten Son, that whoever believes on Him should not perish but have everlasting life."[475] Oh, folks, this is *so* important you understand this![476] When you stand before God, you either stand in your sins and receive the judgment of God, or you stand in Christ and receive the mercy of God. And the way to partake of God's mercy is to repent and put your faith in Him who died for you and rose again on the third day. So, what will you do? [Heckler: "I don't know."] She doesn't know.

A number of years ago in Auckland, New Zealand, a guy about 19 or 20 years old tried his hand at skydiving. It was his first time ever. His friends watched him plummet toward the ground. They watched in horror as his body hit the ground! They thought, *What happened?* They rushed up to what they thought was a dead friend. As they got to him they found that he was still alive. He'd landed on a freshly plowed farmer's field. And as he lay there with one of his leg bones pointed vertical, he looked up at them and said, "Boy, did I blow it!" He left it too late to pull the cord.

Folks, don't blow it for your eternity. "Today if you hear his voice don't harden your heart."[477] Cry out, "God, forgive me. I'm a sinner." Confess and forsake your sins, and the Bible promises you will pass from death to life. God will forgive you. He'll reveal Himself to you. You won't hear thunder or lightning or see visions. He'll just make you a brand new person on the inside, and the things you now love, you'll begin to hate. The things you once hated, you'll begin to love. The name of Jesus will become precious to you. The Bible will become alive to you. You'll begin to love Christians, and the Bible says, "You'll know that you passed from death to life."[478]

Folks, when God does something He does a good job. Remember when you were born the first time? It was kind of radical. You didn't exist, and then you did. The same thing happens spiritually. God will transform you on the inside. Death will lose its sting; the grave will lose its victory.[479] Folks, please get right with God today. You may not have tomorrow. Every single year in America 42,000 people are killed in car accidents—42,000! Can you imagine the heap of people? That many people, 42,000—that's how many people die in car accidents. The next time you get in a car you may be getting into a coffin.

Every day, 1,500 people die of cancer. Every single day! Nearly a million people die of heart disease every day. You don't know when your heart's giving out. You can be in eternity in a heartbeat! And every single day worldwide 150,000 people die. They pass from time into eternity, and there's no hope outside of Jesus Christ. He is "the way, the truth and the life. No man comes to the Father but by Me."[480] "Never a man spoke like this man."[481] He said, "I am the resurrection and the life, he that

---

475 John 3:16

476 Once again, labor the cross, like a doctor who has just diagnosed a disease. Now he leans forward, looks the patient in the eye, and with a cure in his hand he personally and laboriously goes over the instructions with the patient. Because if the patient doesn't understand and apply the instructions, he will die.

477 Hebrews 3:7,8

478 1 John 3:14

479 See 1 Corinthians 15:55–57.

480 John 14:6

481 John 7:46

believes in Me though he dies yet shall he live. Whoever believes in Me shall never die."[482] You have the promise of everlasting life from the God who created all things.

Please take seriously what you've heard today. Call upon the name of the Lord, confess and forsake your sins, pick up a Bible and obey what you read, and God promises He'll reveal Himself to you. Jesus said, "He that has My commands and keeps them, he it is that loves Me. And he that loves Me will be loved by My Father; I too will love him and will reveal Myself to him."[483]

* * *

Tomatoes and Classical Music

A number of years ago I turned the television on to see a sight I could hardly believe. It was a grown man playing castanets to his tomatoes. I thought, *Man, what is going on here?* And then I saw something that astounded me even more. This man put *ear phones* on his tomatoes to play classical music to them. I thought, *What's going on? Why is this nut playing music to his fruit?* (I don't know if you know tomatoes are actually a fruit.) I thought, *What is the BBC wasting time on this guy for?*

Then I heard something that changed my mind radically. They said, "This man has his name in the *Guinness Book of Records* for growing the world's largest tomatoes—*four and a half pounds!*"

I learned something that day, and that is: don't judge something until you see its results.

Now, we learn that with a caterpillar. When we see a caterpillar winding itself a new cocoon, we don't say, "The stupid little grub. What are you doing that for?" We know that the caterpillar is making a chrysalis, and we know that inside that chrysalis a miracle of God is taking place. It's called metamorphosis. Then the little caterpillar's legs fall off, he gets brand new legs, and he becomes a colorful butterfly and breaks out of that cocoon and flies.

Now, folks, maybe you look at Christians and say, "Look at those stupid Christians. Look at the bunch of them, winding themselves, tying themselves up in rules and regulations, hiding from the real world in the "cocoon" of Christianity.

Uh, uh. Don't judge it until you see the results. Right now you're looking at the little grub. What you don't realize is that God is working a miracle within me as a believer. A metamorphosis is taking place. The day will come when I will be transformed in a moment, the Bible says "in the twinkling of an eye."[484] I don't know if you realize that God promises every single person who repents and puts

---

482 John 11:25
483 John 14:21
484 1 Corinthians 15:52

their faith in Jesus Christ a brand new body. They're going to partake in the Kingdom of God, when God's Kingdom comes to this earth—when God's will "will be done on this earth as it is in heaven."[485] That means no more hurricanes, no more tornadoes, no more floods, no disease, no dentists, no death, no dandruff—brand-new bodies for all those who trust in Jesus Christ. You say, "Ah, yeah, yeah. Pie in the sky when you die." No. It's steak on a plate while you wait. You can know for sure what I'm saying is true. This is how.

Imagine if I was filthy rich—I mean a billionaire—and I said to you this morning, "You know I feel sorry for you folks, and I kind of like you. So what I'm going to do is I'm going to give you a million dollars at 8:00 a.m. Fourteen minutes from now, I'm going to give you a million dollars each, and as a token of good faith I'm going to give you a hundred thousand dollars right now." And I came along this line and I slapped a hundred thousand-dollar bills (good hard-earned money) into your hands as you sit there, in your hot little hand. As you looked at that money you'd think, *Man, this guy means it! He wouldn't have done this if he didn't mean it. This is a token of good faith. I'm going to be a millionaire in 14 minutes... 13 minutes!*

Now, God has done the same thing. He has promised everlasting life to all those who trust Him. But the Bible tells us He's given a down payment to us—a token of good faith. The Bible calls it the "earnest" of the Spirit. He *seals* the believer with the Holy Spirit[486] so that the believer, the Christian, knows that what God is saying is true. And how does He do that? Well, He makes you a brand-new person on the inside.[487] He begins that metamorphosis, that transforming of you as a person. You become a brand-new person—born again. It happened to me 33 years ago[488] and I'm still astounded at the overnight transformation that took place in my life. Folks, God wants to do the same for you.

Now you've got a problem. Before God seals you with the Holy Spirit, you have got to realize that it's the Spirit of Holiness that you're going to be sealed with; and God, being holy, can't have fellowship with sin. He's like a holy judge, who can't come down and become friends with a heinous criminal who has raped and murdered. No, the judge has got a job to do. He's got to see that justice is done.

The Bible tells us that we're all in trouble with God, that we have a Day of Justice when we're going to face God, and He's going to judge us by a standard you'd hardly believe. It's by the Ten Commandments—by those words you'll hardly believe. You'll understand in a moment.

Do you realize that God sees your thought life—that there's nothing hidden from the eyes of Him to whom we have to give an account?[489] He sees what you do in darkness as pure light, and God considers lust to be adultery. Jesus said, "You've heard it said by them of old, "You shall not commit adultery, but I say to you whoever looks upon a woman to lust after her has committed adultery already with her in his heart."[490] He considers hatred to be murder. "Whoever hates his brother is a murderer."[491] In fact, Jesus said, "If you get angry without cause you're in danger of judgment."[492]

---

485 Matthew 6:10
486 Ephesians 1:13-14
487 See 2 Corinthians 5:17.
488 April 1972
489 Hebrews 4:13
490 Matthew 5:27,28
491 1 John 3:15
492 Matthew 5:22

Let's look at the Tenth Commandment: "You shall not covet."[493] Have you ever desired something that belongs to someone else, someone's girlfriend, someone's house, someone's car, someone's wife? Then you're guilty of violation of the Tenth Commandment. Or the Ninth: "You shall not bear false witness."[494] Have you ever lied? Well, the Bible says, "All liars will have their part in the lake of fire."[495] Now, I don't like saying that; it's not pleasant. But it's the truth. That's what's going to happen to liars. Ever stolen something? Then you're a thief. Now listen to the voice of your conscience. It's a God-given impartial judge in the courtroom of your mind, that little voice that says, "Guilty." Listen to it today because that conscience is a warning that what God says is true,[496] that the standard I'm speaking of is a righteous standard. You know it's wrong to lie and steal. You know it's wrong to commit adultery and kill, and on the Day of Judgment you'll be without excuse.

"You shall not take the name of the Lord your God in vain."[497] If you use God's name as a cussword, well, "The Lord will not hold him guiltless who takes His name in vain." Have you put God first? Have you kept the Sabbath holy, or have you made a god to suit yourself? You've created a god in your own image, and violated the Second of the Ten Commandments.

Can you see that you're in big trouble when you face the Spirit of Holiness on the Day of Judgment? Can you see that you'll be damned if God has His way, if He has justice? The Bible says God's "not willing that any should perish."[498] Now imagine today if you went inside . . . that you've got a court case today; you've broken the law, and instead of saying, "Judge, I'm not a bad person. The cop was wrong! I don't like his attitude," and instead you were just honest with the judge, and you say, "Hey, Judge, I don't know what to say. I'm sorry. I can't pay the ten-thousand-dollar fine. I've got no money. I'm guilty. If you send me to prison, you're doing the right thing. I don't know what else to say." And you look at the judge, and the judge goes quiet for a minute—he's quiet because he's kind of shocked that a human being would be so honest. And then you see him write something. What's going on? He calls for the bailiff, and he hands him a piece of paper. The bailiff comes across to you and says, "You're free to go."

You say, "What are you talking about?" He says, "The judge just paid your fine. Ten thousand dollars in his own hard-earned money." You say, "Wh . . . wh . . . why would he do that?"

Wouldn't that have been an incredible thing? Well, that is what God has done for you. Two thousand years ago, He paid your fine in His life's blood. God became a person in Jesus Christ and gave His life as a sacrifice for the sin of the world. "For God so loved the world that He gave His only begotten Son that whoever believes in Him should not perish but have everlasting life."[499] That's what God offers you. What love God has got for you! What love God has got for justice, that He would go to such an extreme to satisfy the demands of Eternal Justice.

But that's what He did through the suffering death and the resurrection of Jesus Christ. Jesus rose from the dead and defeated death. And now God offers you the gift of eternal life.

---

493 Exodus 20:17
494 Exodus 20:16
495 Revelation 21:8
496 See Romans 2:15.
497 Exodus 20:7
498 2 Peter 3:9
499 John 3:16

So what are you going to do? Are you going to justify yourself and say, "There's plenty of people worse than me; I'm not a bad person"? Hey, God will throw the book at you. Or if you're going to say, "I don't know if I want to receive that mercy that judge has offered me. I don't know if I want to receive it. I'll just think about it." If you remain contentious and argumentative, God will throw the book at you.

"Behold the goodness and the severity of God."[500] Come on. Stop your mouth.[501] Bow your head and say, "God, I don't know what to say. I'm guilty of the crime. Have mercy on me. Oh, God, I'm a sinner." Then God will extend His mercy toward you. The Bible says He "resists the proud and gives grace to the humble."[502] So humble yourself today under the mighty hand of God and the Bible says He'll "exalt you in due time"[503] if you repent and trust the Savior. God will seal you with the Holy Spirit.

You know the truth of what I'm telling you. You know there's hope in this dark shadow of death. The Bible says, "To them that sat in the shadow of death a light has sprung up."[504] Jesus said, "I'm the light of the world. He that follows Me shall not walk in darkness, but shall have the light of life."[505] So repent today. Turn from all sin and put your trust in Jesus Christ as Lord and Savior. And the Bible says the moment you do that you will pass from death into life. Now in a moment my friend[506] is going to come and speak to you very briefly. And we both want you to know we don't particularly like doing this. Although we count it a privilege to speak to you, it's very awkward. We don't want your money. We're not saying join a church. We're here only because we care about you and where you spend eternity. So as he speaks, please bear that in mind. After we've spoken we'll offer you a free CD called "What Hollywood Believes."[507] Please take it. You'll not only find it fascinating, but you'll find it helpful. Thank you for your attention.

* * *

Unique—Just Like Everyone Else

Sheep are very predictable creatures. I don't know if you realize it, but if one sheep jumps, the others tend to imitate it. They're predictable creatures. And the Bible likens you and I to sheep. We're very predictable. It's a very good analogy.

I remember back in my teenage years, I was part of the counterculture. That is, we were anti-establishment, part of the hippie movement. As a sign of rebellion I grew my hair long before it was cool to have long hair. I wore a penny T-shirt—that's a T-shirt with a pocket on the left side. I wore faded blue jeans and brown leather sandals. Why? Because I was an individual. I didn't wear a suit or a tie. I was an individual, doing my own thing, just like 80 million other teenagers dressed identical.

---

500 Romans 11:22
501 See Romans 3:19.
502 James 4:6
503 1 Peter 5:6
504 Matthew 4:16
505 John 8:12
506 Emeal Zwayne
507 Available through www.livingwaters.com

We're very predictable. We tend to imitate one another. That's because we're made of the same stuff. We look differently, sometimes we act differently, but we are predictable in certain circumstances.

Take for instance torture. If you and I were subjected to mental torture, placed in a room with bright lights and loud music day after day after day, we'd react the same way. We'd go to the security of a corner, and get into a fetal position. We'd react the same way.

In the Second World War, there was a group of people in a lifeboat. They were *packed* in a lifeboat. Suddenly two hands came on the side of the lifeboat; some guy was trying to get in, and the people panicked. They said, "Don't let him in! Don't let him in! We're too crowded as it is. If he gets in, this thing's going to overturn and we're all going to die. Don't let him in!" So someone got a bayonet and went *Whack!* on his fingers, and cut off all his fingers. And he sank into the water, and they rowed away from that red, bloodstained water.

When someone looked back, they saw a pathetic sight. There was a piece of rope hanging from the boat, and this man had gotten a hold of that rope with his teeth, and he was holding on for dear life, and they took pity on him, and pulled him into the boat.

We'd react the same way. You may think you're individualistic; you may think you're unique, but you're unique just like everybody else! Within you, if you are sane, is a will to live. You too would take hold of that rope with your teeth, because you too don't want to die.

The Bible tells us how we can live forever. The Scriptures tell us there is a god of this world— small 'g.' Not the God who created all things, but the god of this world—"the spirit that works in the children of disobedience,"[508] the Bible says. The Scriptures give him a name: Satan. He "blinds the minds of them that believe not."[509] He holds them captive to do his will.[510] And he's the one who wants to cut you off from the very source of life. And yet God wants to throw you a rope. I trust today you'll take hold of it by your teeth. That the will to live will kick in, and you'll say, "Hey, I'm going to listen to this guy. I'm going to see what he has to say. I'm going to be open to what the Bible says about how I can obtain everlasting life."

The reason we die is because we've sinned against God. All of us—you, me, the whole of humanity. Something dwells in us that's a rebellious attitude to the God who gave us life, and you can see this in the way people talk. They use God's name as a cussword, in blasphemy. God gives them life. He gives us the day, the sun, the food we eat, the music we listen to. He gives us a brain to think with, He gives us a tongue to speak with, and we use His name as a cussword! That's called blasphemy. Have you ever done that? More than likely you have. It's very common. That's called "sin." It's the breaking of the Third of the Ten Commandments.

Or take for instance lust. The Bible says of Christians: "You have escaped the corruption that's in the world through lust."[511] Drive along the freeway, look at the billboards—you can see lust, reaching out to grab your eyes. Turn on the television, lust is there. Half-clad women, designed to stir up lust

508 Ephesians 2:2
509 2 Corinthians 4:4
510 See 2 Timothy 2:26.
511 2 Peter 1:4

in the human heart; and we love it! We "drink it in like water,"[512] the Bible says. But listen to what Jesus said: "Whoever looks at a woman to lust after her has committed adultery already with her in his heart!"[513] Have you ever done that? Who hasn't! You're guilty of adultery in God's sight. Have you ever lied or stolen? Then you're a lying thief. You've broken the Ninth and the Eighth of the Ten Commandments. Have you ever desired something that belonged to someone else? Then you've broken the Tenth Commandment. Have you ever hated someone? Then the Bible says you're a murderer, because God sees the thoughts. How many times on a freeway have you said to yourself, "Man, I could kill that guy!" That's murder in the heart, and God considers you guilty.

If you as much as get angry without cause, the Bible says you're "in danger of judgment."[514] So how will you do on the Day of Judgment, when you stand before God and give an account of every idle word, and every deed done in darkness? You know you'll be guilty; your conscience tells you you'd be guilty! Who would not be guilty of transgressing that law? Well, listen to the judgment that's coming: the Bible says, "All liars will have their part in the lake of fire."[515] What a terrible thing! You and I don't think lying is wrong, but the Bible says, "Lying lips are an abomination to the Lord."[516] God is holy, He is pure, He is just, and He's righteous—He's not like us. He sees sin from a perspective of perfection. And that's what you've got to face on Judgment Day—a holy God who requires an account of every idle word, and every deed done in darkness. What will you do on that Day? Man, you don't want to go to Hell. I don't want you to go to Hell. I don't want death to seize upon you, and you be taken to Hell for your sins! Something in you should be saying, "Oh, I don't want to die. I don't want to die!" That's your will to live.

You're like a thief who's been stealing in the darkness of the night. The darkness has been his security. Suddenly a police spotlight comes on him. He's exposed. He's undone! Ten police sharp-shooters have his pounding heart in their sights. What should he do? Carry on his deeds in the light? No! His only sensible avenue of escape is to lift his hand in surrender and say, "I surrender," if he wants to live. And *your* darkness has been your security. The blackness of your ignorance has been your security. That's why you continued doing what you knew was wrong. That's why you continued to feed on lust and pornography, and blasphemed the name of the God who gave you life. You thought nobody saw you. But realize today that you're under the spotlight of the omniscience of Almighty God. He sees all. He knows all. Nothing is hidden from the eyes of Him to whom we have to give an account.[517]

Those Ten Commandments that you've violated are like ten great cannons pointing at your pounding heart. Your only avenue of escape is to lift your hands and say, "I surrender. Oh, God, I surrender. I'm a rebel. God, forgive me. I'm a sinner." And God can forgive you, because He's made provision for your sins to be washed away. He sent Jesus Christ—God in human form, a perfect human being—to suffer and die on a cruel cross. And when Jesus was on the cross, He was laying His life down—taking the punishment for your sins and mine. Jesus said, "No man takes my life from Me. I have power to lay it down, and I have power to raise it up."[518] The Bible says, "God commended His

---

512 See Job 15:16.
513 Matthew 5:27,28
514 Matthew 5:22
515 Revelation 21:8
516 Proverbs 12:22
517 Hebrews 4:13
518 John 10:18

love toward us, in that while we were yet sinners, Christ died for us."[519] Do you understand that? That the God who gave you life is not willing that you should perish, but He's made provision for you to be forgiven, so you can live forever, so you can keep that life.

The Bible says, "What should it profit a man if he gains the whole world, and loses his own soul?"[520] Think about it for a minute! You're going to sink into the waters of death and Hell, but God throws you a rope, and by faith you can take a grip of that rope and you can live, because of what Jesus did on the cross in taking the punishment for your sins. Then the Bible says He rose again on the third day and defeated death.

"Oh grave, where is your victory?"[521] the Bible says. It's been swallowed up in Jesus Christ![522] I cannot tell you the joy I have as a Christian in knowing that I have everlasting life. There's no greater joy to pass on to your children than when they come to you and say, "Daddy, why am I going to die? I don't want to die." You don't have to say, like most people, "Oh, it's just natural. It's okay. It happens to everyone" No, it doesn't! The Bible says death is your enemy![523] Did you know that? It's your enemy!

Are you going to run at death and Hell as though it was your friend? No. Come to your senses today. Repent and trust Jesus Christ. Cry out, "God, forgive me. I've sinned against You. I've violated Your Commandments. I'm worthy of Hell. But I thank You that Jesus Christ died in my place, and rose again on the third day. I repent and trust Him as Lord and Savior."

Folks, the moment you do that, He'll remit your sins—He'll forgive you. He'll make you a new person. You'll pass from death into life. Do it today, because you may not have tomorrow. You don't know how long you have on earth. The only thing you can guarantee is the air going into your lungs at the moment. You can't guarantee another breath. That comes by the mercy of God, and He may lose patience with you, saying, "Tonight your soul is required of you."[524] What a fearful thing to happen! So please think seriously about this. Pray before God—say, "God, forgive me." Pick up a Bible. Seek the Lord, obey Him with your whole heart, and He'll reveal Himself to you.[525]

* * *

Roller-Skate in a Buffalo Herd

Many years ago there was a hit song, the words of which went something like this: "You can't roller-skate in a buffalo herd." And then the writer of the song substantiates that great truth by repeating it. It said, "You can't roller-skate in a buffalo herd," and then he says the third time, "You can't roller-skate in a buffalo herd." Then comes the climax of the song: "But you can be happy if you've a mind to."

---

519 Romans 5:8
520 Mark 8:36
521 1 Corinthians 15:58
522 1 Corinthians 15:54
523 1 Corinthians 15:26
524 See Luke 12:20.
525 See John 14:21.

Now I've got two points of contention with the songwriter. Number one: you *can* roller-skate in a buffalo herd. You just need a big flat area, preferably concrete, and the buffalo herd needs to be fairly thin. And secondly, you *cannot* be happy "if you've a mind to." You see, the word "happy" and the

word "happen" seem to come from the same root word. We're happy because of what happens to us. Let me give an example.

A young man is in a car. His car's full of gas, his pocket's full of money, his girl's snuggling up next to him, and the birds are singing. He's happy, because of what's *happening*.

The young man runs out of money, can't afford gas, his girlfriend leaves him, the car doesn't move, the birds aren't singing—he's not happy, because of what's happening to him.

Now the Christian doesn't find his sustenance from "happiness." He finds it from something called "joy." Now before you demean "joy," and say, "Ah, yeah, that's for Sunday school kids," remember, joy is what you live for. Why do you party? Why do you buy things? Because you want to en*joy* yourself; you want to put *joy* within yourself—you want to en*joy* yourself. And the Christian has joy because the Bible says his name is written in heaven. His name is written in the Lamb's book of life.

That means the Christian knows that he's got eternal life. You say, "You can't know that!" Oh, yes, you can. Let me give an example. A little kid's looking at a heater. His dad comes in and says, "Son, that heater is hot; don't touch it." The kid says, "Okay, okay…" He's just four years old.

Dad goes out of the room. The kid thinks to himself, *I wonder if that heater really is hot.* So he reaches out his little hand to grab that orange heater bar, and as he does, his flesh burns—*Sssssss.* He stops *believing* the heater's hot; he now *knows* it's hot. He's moved out of the realm of *belief* into the realm of *experience*.

Before I was a Christian, I believed in God's existence. I mean, who doesn't? Creation proves there's a Creator. You cannot have a *creation* without a *Creator*. I believed in Jesus Christ. I believed He was the Son of God, that He rose from the dead, that He died on a cross. I believed all that intellectually. But on the 25th of April 1972, at 1:30 in the morning, I reached out and touched the heater bar of God's love and mercy—*Sssssss*—and stopped *believing*. I came to *know*. There's a big difference between belief and experience.

Now let's go back to the kid and the heater. Imagine if a heater expert comes in and says, "Hey, son, that heater is not hot. I'm an expert. I deal with heaters all the time." The kid's not going to believe what that guy says, because he's not in the realm of *belief*; he's in the realm of experience.

And you can have people come up to you and say, "Ahhh, you're a Christian. No, you can't believe this!" You say, "Nope, there's no way you're going to convince me of anything, because I'm not in the realm of belief, I'm in the realm of experience. I reached out and touched the heater bar of God's love and mercy. I came to know God experientially." That's what a Christian is—someone who knows God. Jesus said, "This is eternal life: that they might know You, the only true God and Jesus Christ whom You have sent."[526]

---

526 John 17:17

There is no way you're going reach out and touch God's love and mercy unless you realize you need it. There's no way you're going to say, "God be merciful to me a sinner," unless you *realize* that you're a sinner. So what I'd like to do today is help you understand that in the same way I came to understand it. I'm going to mention a few of the Ten Commandments to show you what sin is.

If I was to ask you a question this morning, "What do you love most in life?" some of you would say, "Well, I love sex…I love food…I love sports…I love the day…I love living…I love the blueness of the sky…I love hearing birds sing." There're certain things we love. Now remember that. Now years ago when my kids were little, I remember I arrived home one day after I purchased a television set—nice big set for them to watch wholesome cartoons in the afternoon. I opened up the door and said, "Hi, kids. I'm home." My kids didn't even bother to come and greet me. You know what they were doing? They were watching TV. Dad's homecoming, the highlight of their day, had become a non-event.

So I walked up to the television and stood in front of it, and I turned it off and said, "Kids, I got this TV for your pleasure because I love you. But if it comes between you and your love for me, it's going. It's a wrong order of affections. You're loving the gift above the giver."

And, folks, if you and I love *anything* in this life more than we love God, we're loving the gift more than the Giver, and the Bible says that is what's called "inordinate affection"—a wrong order of affections.

The First and greatest Commandment is: "You shall love the Lord your God with all of your heart, mind, soul and strength."[527] Do you love God? He gave you life. He gave you every pleasure that you've got. He gave you eyes to see with, ears to hear with, a brain to think with, taste buds to enjoy good food. He lavished life upon you, and gave you a blue sky. Do you love the God who gave you life?

I know that you don't because you're a human being. We're rebels at heart. The Bible says, "There's none that seek after God."[528] We're unthankful. We're unholy. In fact, we are so against the God who gave us life that we use His name as a cussword to express disgust. Have you ever done that? Have you ever used God's name in vain? That's called "blasphemy" and the Bible says, "The Lord will not hold him guiltless who takes His name in vain."[529]

The Second Commandment says, "You should not make yourself a graven image."[530] You should not make a god to suit yourself—where you create a god you feel comfortable with. Folks, I did that before I was a Christian. I used to pray every night to the god of my own creation—a god I felt comfortable with. And yet the Bible says there is one God. He is perfect. He is holy. He is righteous, and you have to stand before Him and give "an account of every idle word."[531] What will you do on that Day? Listen to what Jesus said about the Seventh Commandment: He said, "You've heard it said by them of old, you shall not commit adultery, but I say to you whoever looks upon a woman to lust after her has committed adultery already with her in his heart."[532]

527 Matthew 22:36–38
528 Romans 3:11
529 Exodus 20:7
530 Exodus 20:4
531 Matthew 12:36
532 Matthew 5:27,28

You see, God sees the thought life. Nothing is hidden from the eyes of Him to whom we have to give an account.[533] He sees hatred as murder. Jesus even said, "If you get angry without cause, you're in danger of judgment."[534] And the Bible says, "Whoever hates his brother is a murderer."[535] How are you doing? Do you realize you need God's forgiveness yet? Do you realize that on the Day of Judgment, if He judges you by those Commandments (which He wrote upon your heart by your conscience), then you'll be condemned? Listen to the Ninth Commandment: "You shall not bear false witness."[536] Have you ever lied? Have you ever borne false witness? The Bible warns, "All liars will have their part in the lake of fire."[537]

The Eighth Commandment: "You shall not steal."[538] Have you ever stolen anything in your life, irrespective of its value? If you have, then you're a thief, and thieves will not inherit the Kingdom of God. Folks, Hell is real! It's God's righteous judgment against crimes. It's God's prison for life—for eternity.

Of course God is going to have a Day of Justice. He's not going to let murderers go away free, or rapists, or thieves, or liars, or adulterers, or fornicators (those who have had sex out of marriage). God will have a Day in which righteousness will be done. What will you do on that Day? So those Commandments show us we need God's forgiveness, and I so thank God that He's rich in mercy, and that He can forgive us because of what Jesus did on the cross. Maybe you've never understood it before. Maybe you've looked at a crucifix or heard that Christ died for our sins, and you've never understood that He was being "bruised for our iniquities."[539] He was being punished for our sins. "Upon Him was laid the sin of us all."[540]

Jesus Christ paid your fine in His life's blood, so you could leave the courtroom on the Day of Judgment. "God commended His love toward us in that while we were yet sinners, Christ died for us."[541] "For God so loved the world that He gave His only begotten Son that whoever believes in Him should not perish, but have everlasting life."[542]

Folks, today—not tomorrow, *today*—call out and say, "God, forgive me. I'm a sinner." Turn from your sins. Repent, and put your trust in Jesus Christ. Not intellectually, but experientially. In the same way you trust a parachute to save you, you trust an elevator to lift you, you trust a taxi driver to drive you. I mean we trust so many things. It's not a matter of *belief*, it's a matter of *experience*. Place your trust in Jesus Christ as your Lord and Savior for your eternal salvation, and folks, you will pass from death to life. Just in the quietness of your heart say, "God, I have sinned against You.[543] Please forgive me." Say, "I trust You, Lord Jesus Christ, as my Savior. I believe You rose from the dead and defeated death, and I yield my life to You." Now if you do that and pick up the Scriptures and obey them, you will find that God will reveal Himself to you,[544] and you too will be able to say, "I *know* Him whom

---

533 Hebrews 4:13
534 Matthew 5:22
535 1 John 3:15
536 Exodus 20:16
537 Revelation 21:8
538 Exodus 20:15
539 Isaiah 53:5
540 Isaiah 53:6
541 Romans 5:8
542 John 3:16
543 See Psalm 51:4.
544 See John 14:21.

to know is life eternal."[545] You too will be able to rejoice because your name is written in Heaven. Folks, there's no greater joy than to know death has lost its sting.

* * *

Judas Was in Charge of the Finances

If I was to approach some of you this morning and say, "Are you a Christian?" one or two of you may say, "Yes." Some of you would say, "Ah . . . I'm not sure; I try to be." One or two of you may say, "I'm not a Christian because there're hypocrites in the Church—that's why I'm not a Christian." And you'd say that because you don't understand what a "hypocrite" is. A hypocrite is not a Christian; a hypocrite's a *pretending* Christian. And the Church is not filled with hypocrites because the Church is not the building; it's made up of the people.

So, as I said, a hypocrite is not a Christian; he's a *pretending* Christian. That's what the word "hypocrite" means. It means "actor." Judas Iscariot was a hypocrite. He was an actor. I've got a friend who used to look at Leonardo Da Vinci's picture of the Last Supper (the painting of the last supper) and he'd say, "What I liked to do as a young man was look for Judas."

Heckler: (Unintelligible) . . . I bet when that tree dies and you'll be long gone, that tree's going to outlive what you are sitting here talking about. [A young man sitting next to him, who seemed to be his son, tries to stop the man from hindering God's Word.] Freedom of speech. If he can talk, I can talk. That tree was put here by God; this guy was not. That tree's not bothering me. That tree's not trying to persuade me.

Son: If you're going to do that . . .

Heckler: Freedom of speech; I can talk.

Son: Hey, did you hear me?

Heckler: I hear you. I'm also hearing something I don't want to hear. I have the right to not be bothered by somebody.

Son: (Unintelligible.)

Heckler: That tree's not bothering me and God put that tree here.

Son: Listen, stop it . . . (Unintelligible.)

Heckler: I don't care. It stopped him for a couple seconds. I am so proud.

Son: Stop it.

Heckler: Do you have permission to record any of this? Because if any of this is used and I ever see it anywhere, I will sue you.

Ray: You cannot see what's been "heard," sir.

Heckler: My, that's a beautiful tree. God made a thing of beauty.

Son: Do you realize you're @#$%! me off?

---

545 See John 17:3.

Ray: Folks, I'd like to agree with this man totally. That tree will be here long after I'm dead and gone.

Heckler: That tree is not bothering me a bit.

Ray: Probably, I've got another 25–30 years, God willing. Life is very transient. The Scriptures say, "Life is but a vapor that appears for a moment and vanishes away."[546] Folks, I would like you to think closely about what this man said. It's so true. Trees last longer than most of us. That big tree... I don't know, it may last 100–150 years. I don't know. But if you're lucky you'll probably average 70 years. Do you know how many people die on roads every year in the U.S.? Forty-two thousand. Every day 1,500 people die of cancer—every single day in the U.S. Just under a million die of heart disease in the United States each year. Throughout the world 150,000 people die every 24 hours. So think about your mortality. This man has got perception. He perceives that life is transient. So while I'm speaking, please think about that.

Now, I was talking about hypocrites before, saying my friend would look at Leonardo Da Vinci's painting of the Last Supper and look for Judas. He'd say, "Where is that evil guy with his hands clasped, hooked nose, counting the money?" That's totally unbiblical. Judas is probably a good-looking guy like my friend here.

He was in charge of the finances, the Bible tells us.[547] Did you know that? He was such a good guy, when some woman bought some expensive perfume and gave it to Jesus, he said, "Why wasn't this perfume sold and the money given to the poor?"[548] He was a good guy, Judas Iscariot. But the Bible gives us insight. It says, not that he cared for the poor but he looked after the finances, and would steal from the finances.[549]

Judas was a hypocrite, he was a pretender, and the Bible says Jesus knew "from the beginning who would betray Him."[550] He knew who was going to betray Him. He knew Judas's heart. He was such a good actor, when he was about to betray Jesus, Jesus said, "One of you will betray Me," and the disciples, instead of looking at old hook-nose down the end and saying, "Ah, yeah, we know what you're talking about," they said, "Is it I, Lord? Is it me?"[551] They suspected themselves rather than the trustworthy treasurer. And even when Jesus said, "It's he who puts his hand in the dish" and Judas went out and betrayed Jesus, the disciples thought that he had gone to give money to the poor. That's what a good actor Judas was.

Oh, yeah, there is such a thing as a hypocrite. But listen to what the Bible says of those who judge hypocrites. In the book of Romans it says, "You who judge another and do the same things, do you think you'll escape the judgment of God?"[552] You see, maybe you see someone who says, "I'm a Christian," but they watch dirty movies, and you know that's wrong. Uh, huh... Do you watch dirty movies? Or you say, "He calls

546 James 4:14
547 See John 12:6.
548 John 12:5
549 John 12:6
550 John 6:64
551 Matthew 26:22
552 Romans 2:3

himself a Christian, but he's a liar." Do you lie? "He calls himself a Christian, but he doesn't pay his taxes—I know that." Do you pay your taxes? Do you think you'll escape the judgment of God if you lie and steal?

Do you realize that God considers lust to be the same as adultery? Jesus said, "Whoever looks upon a woman to lust after her has committed adultery already with her in his heart."[553] Folks, every man will give an account himself to God. I won't give an account of this man. He won't give an account of me. We have to give an account of ourselves to God. And the standard that God will judge us by is the Ten Commandments, and who of us can say, "I've kept that Law?"

Heckler: (Unintelligible discussion with his son.)

Ray: The Bible says if you hate someone, you're a murderer.

Heckler: I am not a murderer… (Unintelligible.)

Ray: Think about it. If you lie or steal, then you're a lying thief. Let me tell you what's going to happen to hypocrites. You'll find this very interesting.

I had a friend years ago who had a little parakeet. They called it "Bluey." It was trained to come and sit on the plate at meal times and eat from the meal. Neat little bird, Bluey. One day this young lady came into the kitchen and left the door open. The bird saw the blue sky and thought, "Yippee, freedom!" and it flew toward the door. The girl loved the bird, so she ran to the door and slammed it … and stopped half that bird getting out.

Folks, if you read the book of Matthew, chapter 24, the same thing will happen to the hypocrite. The Bible says, "If that servant shall say in his heart, 'My Lord delays his coming,' and begins to smite his fellow servants and drink with the drunkard, the Lord of that servant will come in an hour he thinks not and cut him in half and give him his portion with the hypocrites."[554] So, folks, if you know hypocrites aren't real in their faith, how much more does Almighty God? And He'll hold them accountable. The Bible makes it clear, all hypocrites will go to Hell. Folks, I don't want you to go there either. If you've got a conscience this morning, you'll know you've violated those Commandments. I mean, is God first in your life? Do you love the One who gave you life—with heart, mind, soul, and strength? A man who lacks understanding will look at a tree and see a tree. When a man who has insight looks at a tree, he'll not just see a tree, he'll see it as the Creator's hand—the genius of Almighty God.

Folks, when you become a Christian, God will open the eyes of your understanding. No longer will you just see a sunset, but you'll see it as the creation of Almighty God … the blueness of the sky … you'll see birds fly—all things will become new. All things can become new for you today because of what God did through Jesus Christ 2,000 years ago. All of us have violated those Commandments. None of us can point a finger at the other. We've "all sinned and come short of the glory of God"[555]—every single one of us, and the Bible says, "All liars will have their part in the lake of fire."[556] Hell is a reality. God will have His Day of Justice, and if you've sinned when you stand before God on the Day of Judgment, Hell will be your portion for eternity. I don't want that to happen to you. You don't want it to happen to you. That's not God's will. He made provision for you to be forgiven through the suffering death and the resurrection of Jesus Christ. When Jesus was on the cross,

553 Matthew 5:27,28
554 Matthew 24:48–51
555 Romans 3:23
556 Revelation 21:8

He took the punishment for the sin of the world. All your sins were laid upon Jesus Christ. He paid your fine in His life's blood. "For God so loved the world that He gave His only begotten Son that whoever believes upon Him should not perish but have everlasting life."[557]

Folks, that's what God offers you through Jesus Christ—everlasting life. So what must you do to find that salvation? How can you find God's forgiveness? Repent—an old-fashioned word, it just means to turn from your sins. Confess and forsake them. Trust Jesus Christ. Folks, don't just *believe* in Jesus; *trust* in Him. I believed in Jesus before I was a Christian. The Bible says "demons believe and tremble."[558] You must trust in the Savior, and the moment you do that, the "eyes of your understanding will be enlightened."[559] You'll be like a man born blind, who suddenly sees light. The Bible will come alive. You'll get a love for other Christians. Creation will never look the same. You'll see the genius of God's creative hand—even in an ant.

Salvation...when God comes to you and transforms your life, folks it's so real. Remember the first time you were born? You did not exist, and suddenly you did. That was radical! The same thing happens when you're born again. You'll receive a new heart with new desires. When you become a Christian, something terrible happens...and this is what it is. What you do is you realize that Heaven is a reality, and suddenly you realize that Hell is also reality. You realize the Bible is true when it says most will put their hands over their ears and rush at Hell as though it were Heaven. And that's horrific, folks. That's why I'm here today, to say God can forgive you. He can wash you; He can cleanse you.

Ray: How you doin'?

Police Officer: Okay. Can I talk to you a second?

Ray: Yeah, sure. Do I have to move from here?

Police Officer: Yeah. I need to talk to you about that...

* * *

---

557 John 3:16
558 James 2:19
559 Ephesians 1:18

92

EMEAL ZWAYNE ("EZ")

Introduction

HI, my name is Emeal Zwayne, and I have the privilege of serving as the General Manager of Living Waters Publications and The Way of the Master. Perhaps you know me better by my initials, E.Z.

I trust that you've enjoyed the open-air preaching segments by Ray Comfort. Ray and I have had the privilege of preaching together on a daily basis for the last two-and-a-half years at a local courthouse in Bellflower, California. This has been an absolute thrill for us because previously, Ray and I have preached together for three-and-a-half years every week at the Third Street Promenade in Santa Monica. This is a long strip filled with shops on either side, passersby, and shoppers, and performers and, as you can imagine, hecklers as well. Ray and I used to constantly compete with the noise of the music from the performers and the loud screams of the hecklers that would be passing by. And so, you can imagine how overwhelmed with joy we were when we discovered that God blessed us with a place to preach every day, where the audience was not only a captive audience but usually a congenial one as well.

You'll notice in the following segments that I don't typically go deep into the Law while I'm preaching. This is because Ray and I are tag-teaming in a sense. Ray preaches first. He expounds the Law. He shares the gospel. And then I follow him and share, for the most part, an illustration that will capture everything that Ray had just shared. So you'll notice that as I speak I typically share illustrations that will make what Ray and I have been talking about memorable in the minds of the listeners. That's done intentionally so that when they walk away they can leave with the gospel encapsulated in their mind.

This has been an absolute joy for us to preach the gospel in the open-air to people who are not only listening, but are also allowing us to give them tracts, books, and CDs. And we walk away oftentimes praising God for this opportunity that He's given us. I remember there have been times where I felt a sense of apprehension as we've approached the crowd. And on one particular occasion I thought of an individual who's drowning, and the moral obligation that I have as a decent human being to save them. Now in that case, the last thing that I would need is the permission of the drowning man. That drowning man is in a desperate situation. He may reject my help, but I'm obligated nonetheless to give it. Oftentimes we feel a sense of apprehension as we're approaching because we feel as if though we need everyone to give us their permission or approval to even preach. But then I've come to realize as well that the God of the Universe who created the very plot of land that we stand on is the One who gave us, not only the permission, but the command, as He said, "Go into all the world"—His world, His earth—"and preach the gospel to every creature."

It's my prayer that you will be encouraged through what you're about to hear, that your heart will be stirred to go forth and proclaim the everlasting gospel as an ambassador of Jesus Christ. That you would allow God to plead through you to this world that they would be reconciled to God on behalf of Christ.

Cutting the Cord

Folks, we thank you so much for giving us your attention this morning. My friend and I have been doing this now on a daily basis for almost two-and-a-half years, and really words can't even begin to express how much we appreciate your patience with us. Over the course of the last couple of years we've occasionally had someone yell something out from the crowd while we're speaking. And perhaps the most popular comment we've heard is, "Get a job!" I'm sure that's crossed some of your minds this morning. I want to assure you we have jobs. The gentleman who just spoke to you is a best-selling author who has written over 40 books. He hosts an international television program with the Hollywood celebrity Kirk Cameron, from the "Growing Pains" sitcom and the *Left Behind* movies. It airs in over 70 countries around the world with millions of viewers. And both he and I are executive officers of a multi-million dollar corporation. And to be honest with you, we'd much rather be hiding out in our nice comfortable offices this morning but as my friend said, "We're here because we care." Because we know that there's a real place called Hell, and if any of you standing in this line die without having received the forgiveness that God has provided through His Son, friends, you'll spend an eternity there and we don't want that for you.

Now I know that there's a lot that's going through your minds this morning. I know some of you are saying, "You know what, I've tried this Christianity thing. I've tried this spiritual journey thing and it's never gotten me anywhere." There's a reason for that. There's an interesting story about these two men who lived on an island off the South Pacific and every weekend they were accustomed to going island hopping. They'd hear about these big parties and so they'd take their little boat and go from island to island and party all night long.

Well, one day they hear about this real big rager that was going on, on a nearby island, and so they got ready—they got dressed up and loaded all their kegs and booze on their little boat. They hopped in and they rowed to the island (it took them about an hour). They got out of their boat. They tied the boat to the dock. They went on the island. They partied all night long.

About one o'clock in the morning, they made their way back to their boat, stumbling all over the place, practically passing out. They made it to their boat, they hopped in. They began to row back to their island. They rowed and rowed and rowed. And you know, they're singing their drunk men songs, and before long as time had passed, they looked out toward the horizon a little more sobered up. They wiped their eyes. They looked again. They looked at each other. They couldn't believe it! The sun was beginning to rise. They thought to themselves, *Wait a minute! It only took us about a hour to get to this island. We left about one o'clock in the morning. We've been rowing for five, six hours.* They looked to the right, there was no sign of land. They looked to the left, there was no sign of land. They looked in front of them, there was no sign of land. It wasn't until they turned around that they realized that they hadn't moved an inch because the boat was still tied to the dock.

And friends, that's what it's like. We're trying to make our spiritual journey. We're trying to get somewhere for God, but we look and we're in the same place. We're expending energy. We're moving. We're doing all this stuff, but we're in the same place. Those guys could have rowed until the oars rotted in the water, until their arms fell off. But until they untied that boat from the dock, they weren't going to get anywhere.

And friends, as long as you haven't repented of your sins, and your heart is still tied to your sin that you so love, no matter how much energy you expend spiritually, how much you go to church, or pray or read your Bible or try to do good things, until you repent you'll never move out of darkness into light. You'll never receive forgiveness of sin and you'll never have a relationship with God, and you'll never see eternal life in Heaven.

I want to urge you this morning to acknowledge that God is extending love and mercy and grace toward you. That's the message of hope we're sharing. We shared the bad news with you—that you're guilty, you've broken God's Law, you're deserving of His judgment. But we want to leave you with the "Good News." The Bible says, "For God so loved the world that He gave His only begotten Son that whoever believes in Him should not perish, but have everlasting life."[560] We want to urge you today to acknowledge that you've sinned against God, and that Christ died on the cross for your sins, and rose again on the third day. And if you repent, God will grant you everlasting life and forgive you of every sin you've ever committed.

So please consider that today, and folks, realize that what you're hearing today is by no accident. God is trying to get your attention. He's granting you another offer of mercy and grace. We thank you so much. We so respect your attention. We know you didn't have to so much as look our way. As we come along with one of these CDs, if you'd like one, please feel free to take one—it's free of charge. It's called "What Hollywood Believes" and it talks about the intimate beliefs of Hollywood celebrities. So thank you so much, and God bless you.

* * *

Grasping His Greatness

I know that some of you listening to us this morning have several things going through your minds. Some of you are a bit angry at God—you've got a bone to pick. And you've said in your mind, "You know what? When I stand in front of God on that day, I've got a few words for Him." And that attitude is quite prevalent today. A number of people are beginning to feel that way. They're going to have a few words for God when they stand in front of Him. I think the reason for that attitude is because many of us have really forgotten who God really is, and in the light of that, we've forgotten who we are. I'd like to put that in perspective for you this morning.

You know if a beam of light came shooting through this place right now at the speed of light, 186,000 miles per second, and we happen to have our saddle handy, we jumped on this beam of light and started flying through space at that speed, in a second-and-a-half we'd be at the moon. In nine minutes we would reach the sun ninety-three million miles away. In four years we would finally reach

560 John 3:16

95

Alpha Centauri, which is the star closest to our solar system. But get this: if we started at that same speed, 186,000 miles per second and went non-stop starting at the beginning of our galaxy, the Milky Way, and keep going non-stop, it would take us 100,000 years to get from one end of our galaxy to the other. Now get this: there are over 100 billion galaxies with over 200 billion stars in each galaxy, and the Bible tells us about God that heaven and the heaven of heavens can't contain Him. And that He spans the universe with His hand. This is the God who created us. What are we in comparison to the scope of the universe? We're less than microscopic.

When we stand in front of God on the Day of Judgment, we will have nothing to say. It's my hope that you and I will say something before we reach that Day, and the only thing that's going to get us into God's kingdom is an admission of our guilt, a confession of our sin, and a faith in the death and resurrection of His Son whom God sent because of His love for us. So, folks, I urge you to turn from your sin, place your faith in Christ, and be safe from the wrath to come.

Thank you again so much for your attention. If you'd like one of these CDs, please take one. It's called "What Hollywood Believes" and it will tell you about the intimate beliefs of Hollywood celebrities, and it's free. God bless you.

* * *

He Healed Our Wounds

I'm here because we know that there's a real place called Hell, and if any of you standing in this line die without having received the loving and gracious forgiveness that God has provided through His Son, then you'll spend an eternity there. And honestly, folks, that's a thought that we just can't bear. I know that some of you listening to us this morning might think that what we're doing is a bit arrogant. You might perceive this as talking down to you or speaking from a vantage point of being "holier than thou." I want to assure you that's not the case.

There's an interesting story of a man who was driving down the street one day and found a wounded dog on the side of the road. He had compassion and mercy, so he picked up the dog, put it in his car, drove it home. He brought it in and for days and days he took care of it. He bandaged its wounds. He feed it. He nursed it back to health, gave it affection and attention, and the dog made a full recovery. And on that final day when he took off the final bandage and put the dog on the ground, it immediately scampered off and ran out the door.

The man stood there and he thought, *What an ungrateful little creature! I mean, I took it off the side of the road. It was on the brink of death. I bandaged its wounds, I nursed it back to health. I feed it. I gave it attention and affection, and look at its response.*

A few hours later, the man heard a scratch on his the door. He opened the door and he found that dog with another wounded dog standing right beside it.

You know, folks, that's all we're doing here today. God had mercy and grace on our souls. He opened our eyes and He gave us eternal life, and we're just here today telling you where you can find that same eternal life: in the hands of a loving Creator, a merciful and gracious God who loved us so much that two thousand years ago, He became a man for our sake when He wasn't obligated to do it. No one forced Him to do it. In fact, all He owed us was wrath and judgment because of our wickedness. But in His love and mercy He hung upon that cross and rose again on the third day, and now He offers you forgiveness and everlasting life, if you repent—if you acknowledge what we've shared with you today.

With God, when you deny you're guilty, you not only remain guilty but you also incur greater guilt. But when you agree with God, when you admit that what He says is true—that you are sinful, that you've violated His holy Law, that you're deserving of His judgment and then acknowledge that He has made a way for you to be saved—then He makes you innocent. He wipes away all your sins. He moves that burden that weighs you down and He gives you everlasting life.

* * *

Lord, Lord

We live in the United States of America, which is considered by the whole world as being a "Christian nation." We're as much a Christian nation in the eyes of the world as Saudi Arabia is a Muslim nation. And there's a bit of a fear in that, because when that happens to a society, everyone becomes in a sense convinced of the fact that they are a genuine Christian, according to the definition that Jesus Christ gave. But there's a passage of Scripture that has really struck a chord of sobriety in my own heart. It's Matthew 7:21–23. Jesus said, "Many will say to Me on that Day, 'Lord, Lord, have we not prophesied in Your name, cast out demons in Your name, done many wonders in Your name?' He said, "And then I will say to them, 'I never knew you. Depart from Me, you who practice lawlessness.'"

In other words, what Jesus is saying is that there will be a group of people who will have lived their lives on earth, the whole time having thought that they were genuinely Christians, but in the end they will die realizing that God never even knew them. I mean, that's *really* sobering. The reason for that is because many of us think that what we do on the outside in terms of Christian rituals, and duties, and obligations—that genuinely makes us a Christian. But that's not the case.

If you and I go down right now to the local Oscar Mayer pig farm, and we pick up a nice little pig—we take it home, stick it in our bathtub, pull out the Epi-lady and the Bic razor, shave it down, pour gallons of perfume and cologne on this little pig, take it to the local tuxedo shop, get it a custom-made pig tuxedo with a bowtie, a top hat, the whole deal—take it to get it some reconstructive plastic surgery, take it to the dentist, and get it some nice dentures made. Then we take this pig home, we put it in front of a banquet table, we fill this table with all the greatest delicacies you can imagine. We invite our guests of honor.

Let me ask you a question: When you let that pig go, what's the first thing that that pig will do? Is it going to be concerned about being adorned with the products of fluff? The first thing that pig is going to do if you give it a chance is it's going to run right back to the pigpen. The reason for that is because it's a pig at heart. It doesn't matter what you do to its outside. It doesn't matter how you change it and reconstruct its face and make it look like a person—it's a pig at heart. And that's kind of what it's like with many of us who are trying to live the Christian life, but who haven't had a change of heart. It doesn't matter how much we sit in church, how much we pray, how much we try to look like a Christian—until our heart has been changed, it doesn't make a difference. We're going to always go right back to the pigpen because that's where our heart is.

A lot of you today are saying, "You know what? I've tried this Christian thing—I've gone to church, I've gone to these big events where I've said a prayer and whatnot. But you don't understand—that's just not me. I don't appreciate or really enjoy the things of Christianity." Well, my answer to you is, "Don't be surprised."

If you and I go right now and we find a little caterpillar crawling around in the grass, we stick it in a little caterpillar box, complete with all the great features that a caterpillar would enjoy. We get a plane ticket, fly over to New York, take a cab downtown, go into the elevator of the Empire State Building, go up to the top, take out our jet-lagged little caterpillar, stick a nice little tight blue suit on it with a red cape, stick a little 'S' on its chest then dangle it over the edge of the Empire State Building and, just in case any of you are Insect's Right's Activists, "accidentally" let it go, then go back down to ground level, to what's now *splat* all over the sidewalk, raise this caterpillar from the dead, and ask it if it had a pleasurable experience being tossed from the top of the Empire State Building... It's kind of a ridiculous question. Of course not!

But you take that same creature, that same entity after it's been transformed into a butterfly. You take it to the top of the Empire State Building and let it go, then interview it afterward and ask it if it had a pleasurable experience, and you're going to get an entirely different response. Of course it did. It's the same creature, the same entity, but something has happened. After it went through that metamorphosis and it was transformed, that which at one time was unnatural for it has become the most natural and enjoyable thing.

And that's why Jesus Christ said, "You must be born again and unless one is born again, you cannot enter the kingdom of God."[561] And that means you must be born from above. You must receive a new nature. Your heart must change and as my friend shared, that happens through repentance—by acknowledging that you've sinned against a holy God, you've offended Him, and then realizing the seriousness of your sin, and then understanding the love of God that was demonstrated through the crucifixion and the resurrection of His Son.

So, folks, please consider that today, and understand that we're sharing this with you again because we care. I know that many of us gamble in our hearts and minds and we think that we're going to live another day, and maybe we'll get an opportunity to make it right with God. We're not promised tomorrow, my friends. I want to urge you to recognize the grace that God's extending to

561 John 3:3

you today—that He's letting you hear the precious truth of His gospel, that He wants to give you the free gift of everlasting life. You can't earn it, you can't work for it, you don't deserve it, but He's willing to give it to you freely if you humble yourself and repent and turn from your sin. So, folks, thank you so much for giving us your attention this morning again. We know you didn't even have to so much as look our way, and as you lay your head down on your pillow tonight, remember that you've heard the truth and it's by no accident. What you do with it is up to you.

* * *

On the Brink of Eternity

I've come to realize that we live in a country that is filled with distraction from that which is most important. Think about it; think about how lightly we look at the reality of death. We joke about it, we have various sayings in relation to it like, "Oh, man, I feel like I'm going to die!" Or, "I'm going to kill you!" or things of that sort. We make light of it; we joke about it because it's almost overwhelming to really focus on. And you know there's a good correlation to that.

We're sitting here right now, you look at the sky—we don't see a cloud, the wind is hardly blowing. But on the other side of the country, connected to the same piece of land right now is a hurricane that is absolutely devastating people. There are ten thousand people in one location in that dome there in Louisiana. There are people hiding in their attics right now, winds tearing roofs off their houses. It's devastating, but the furthest thing from our mind. You sit here and you look around, and it's so calm and peaceful and quiet. Who would think that there's a storm connected to this land mass? Friends, that's so similar to what's taking place in the eternal realm. You know that you and I are literally on the brink of eternity, and many of you are right now on the brink of God's wrath and judgment for all eternity. I don't enjoy talking about Hell. I don't do it because it's fun. If it were up to me Hell wouldn't exist, but it's not up to me. Hell is a real place and we're all in danger of it, if we don't have the salvation that God has provided through His Son.

* * *

Setting the Standard

I know that all of you were glued to your television screens in the last few days because of all the different events that were going on in the news. And one of those events were dealings with a "B.T.K. Killer"—the man from Kansas who dubbed himself the "B.T.K. Killer," meaning "Bind, Torture, Kill," who's been convicted of killing ten human beings. I remember watching the proceedings as he was being sentenced as he confessed his crimes. I remember listening to some of these victims, the families of some of these victims and the things that they were saying. I remember logging on to CNN.com and seeing the heading. It said, "Families to the B.T.K Killer." Their message to him was, "Rot in Hell!" I remember hearing one man saying how he wished that he would develop some disease and slowly suffer and then choke on his own vomit, and die that way. Just hearing all the things—people calling him a "cesspool," and human waste, and just all these things. And I just thought of that, and I thought, *Man, look at that. These people want him to suffer. That human sense of justice for the things that he had done.* I thought of that, and then I thought about how oftentimes, we as people, when we speak about God we just talk about Him as being one who'll just overlook the things that people have done. How He'll just kind of forget about it and pass it over. And I thought, *Can you imagine, if that judge yesterday in that courtroom looked at that man and said, "You know what? I know that you've done all these things, but I'm a caring, loving, forgiving judge, and I'm just going to let you go today."*

Can you imagine the horror that would have overtaken all of America and the world, because we have a sense of justice? But it's amazing that when we look at ourselves in light of God, we think that He should just overlook all that we've done. God is a God of justice. The problem is that we compare ourselves to someone like the "B.T.K. Killer" and we don't look so bad. We compare ourselves to someone like Hitler and we don't look so bad. But, friends, on the Day of Judgment, God is going to compare us to His own standard of righteousness—as my friend said, His Commandments—when we compare ourselves to that, then we see ourselves in the true light. We realize why we deserve God's wrath and judgment, because we've broken His Commandments.

If you see a white sheep eating green grass, it doesn't look so bad. It looks so white and pure. But you take it into a barn, let it snow overnight, bring it out, and you look at this sheep now with this freshly fallen, pure white, snowy background, and it doesn't look so white anymore.

When we compare ourselves to God's holiness, we see ourselves in our true light. If we've lied, by His standard, we're liars. If we've looked with lust, we're adulterers. The Bible said if we've had unjust anger or hatred toward anyone in our heart, we're guilty of murder. And by that standard, suddenly we see ourselves deserving of God's judgment, and to think otherwise would be an insult to God. That's why we need to recognize today our need for God's forgiveness.

There are degrees of justice and judgment. If I come up to you today and I run toward you and your friends with you, and they pull out Uzis and spray me and kill me, you'd be in big trouble. However, if you attempt to do the same thing to the President of the United States, the punishment for your crime would be a lot harsher.

We've sinned against Holy God. We deserve Hell. We need His forgiveness. So I urge you to place your faith in the death and resurrection of Christ today and receive God's grace and forgiveness. Thank you so much for your attention.

* * *

The Invincibility Complex

I've often been amazed at the fact that even though the majority of Americans claim to believe in the existence of God, and in the reality of Heaven and Hell, that very few of us do anything significant about it. I came to realize years ago it's because many of us have what I call "The Invincibility Complex." We don't really think that day will ever come about when we'll die and stand before God and give account. This sort of deceptive mental block kicks in subconsciously and we begin to think ourselves to be untouchable—invincible!

I remember having this mindset as a child. I remember one time riding on a bus, going on a field trip with my class, when all of a sudden I was overwhelmed with this sense of invincibility, and I looked over at my teacher. I said to her, "You know what? I have this feeling as if I'll never die—that if this bus were to roll off the cliff right now and blow up, nothing would happen to me. I'd just get up and walk out." It was a real feeling; I mean I really sensed it.

It was as if I blinked, and it seemed like just a few short years after that where I was holding my own dear mother in my arms. I watched her turn into eighty pounds of skin and bones and take her last breath as I held her, and she died of cancer. It was a real wake-up call to me, a real reminder of my own mortality.

Then I began to see the modern-day "untouchables" of our day and age, the modern-day super-heroes. I remember seeing Mohammed Ali on my television screen, as he was dancing around the ring, pulverizing his opponents into dust, chanting all his famous sayings: "I'm a bad man! I'm the king of the world!" Again, it was as though I blinked. Sometime ago, I saw the same man standing on a stage to receive an award, and those hands of his that he used to launch like satellite-guided missiles at his opponents were now shaking uncontrollably from the advanced stages of Parkinson's Disease. Those fast-firing one-liners were reduced to slowly spoken two- or three-word sentences. Those feet of his that used to dance beautifully around the ring could now hardly support the weight of his body; and I looked at the "before" and the "after" picture and I was reminded of the frailty and the mortality of man.

I remember seeing another one of our modern-day superheroes. He was a paragon of super-human strength—faster than a speeding bullet, more powerful than a locomotive, able to leap tall buildings in a single bound; and again, I blinked and there was Superman on my television screen—Christopher Reeve, a quadriplegic. He exchanged his cape for a wheelchair, his mighty wind-gushing lungs for a ventilator, his bullet-proof and robust body for one that no longer worked.

Then, recently, we saw this man who we all imagined in our mind to be beyond death actually taken by the cold grip of death. It was a wake-up call to me again. And folks, I want to remind you this morning of *your* mortality—that you are one breath away from standing before your Maker. And the question is: Will you enter His kingdom or will you suffer the just judgment for your sin, which is an eternity in Hell?

I wonder how many of the two-hundred thousand plus precious souls who perished in the tsunamis considered when they watched the events of 9/11 unfold that they would be swept away in a tragedy. I wonder how many of these precious souls off the Gulf Coast of Mexico who watched the events of the tsunami unfold, considered that they themselves would be swept away in a similar tragedy.

We don't plan our deaths. One hundred and fifty thousand human beings die every twenty-four hours, and the question is: Are we good enough to enter into God's Kingdom?

Folks, I'll leave you with this final question this morning: What is it that you hope in? God sent His beloved Son to hang upon a cross and suffer for your sins and mine because we broke His holy Law, and are deserving of His wrath and judgment. Without the shedding of blood, the Bible says, there is no forgiveness—there is no remission of sin. And then He rose again on the third day, and so in light of that, what do you hope in? And I'll leave you with this poem that I wrote years ago:

> "What is it that you hope in? Is it riches, is it fame?
> To climb the corporate ladder so that all would know your name?
> Do you hope to be the toughest, the buffest of them all?
> The biggest, baddest giant that'll make the rest look small?
>
> Do you hope to rule the world and conquer every land?
> That all would be your servants and your wish is their command?
> Is your hope in your appearance, in the apparel that you wear?
> To have the finest wardrobe and the picture-perfect hair?
>
> Is your hope to gain possessions, or find the perfect mate?
> Or to be the most artistic and to wondrously create?
> Is your hope in finding pleasure of every sort and kind?
> Or seeking after knowledge just to satisfy your mind?
>
> What is it that you hope in? What will you gain or earn?
> I tell you if it's temporal then it's guaranteed to burn.
> I put my hope in Jesus and for Him alone I yearn
> He is my hope forever, He alone is my concern."

And, friends, I hope that that will become true for you as well; that today you realize that God is giving you an opportunity of grace—that you have still the breath of life within you, but I urge you not to harden your heart. I urge you to consider that this could be your last day on the earth. I thank you again so much for your attention here this morning. I know that you didn't have to so much as look my way.

* * *

The Medium

I know some of you are standing in this line here this morning and you're just shaking your head. To you the reality of God is, "Uh, really not a reality at all." In fact, to you it's fantasy, it's mythological, it's a fairy tale. And you just can't come to grips with the fact that God is real, and you rationalize that because you've never seen God. You've never felt God. You've never experienced God in the sense with your senses—never heard His voice, and so you discredit His existence because of that.

Now I want to draw an analogy for you; maybe you can relate to this. Imagine you had a task given to you to go to a primitive tribe living on a distant island. Your task was to convince them of the existence of television and radio airwaves. That was your task. You're going to go and tell these primitive people, who've never seen a television set, never seen a radio in their life, that radio and television airwaves are real, and in fact are all around them. They're right there. They're standing in the midst of them. There're people, voices going through them. There're comedians cracking jokes. There're news men and news women telling jokes in distant parts of the world.

Now imagine what that task would be like. Imagine what would be your first resort. I mean, what would you—what would you want to use to convince these people that radio and television airwaves are real? You're going to tell them, "Listen, I have a device... if you're willing to come with me so that I can show it to you. It's on the other side of the island. You'll be able to see that these things are real." They're going to laugh at you, they're going to mock you, they're going to ridicule you. They're going to think that you're crazy—"What are you talking about, 'People all around me and images'—that's ridiculous!" And unless those people are willing, unless they're willing to look at the medium that will reveal to them the reality of these radio and television airwaves, they are never going to know their reality. Now think about that. I mean, in a sense this is real. You and I know beyond the shadow of a doubt that radio and television airwaves are real, and the reason we know is because we have the instruments that manifest their reality.

Folks, it's the same thing in the spiritual. Look, you can't see radio or television airwaves in and of themselves in their true essence. You can't see them or hear them or taste them or touch them. They're not experienced by our senses unless we have the medium that reveals them, and it's the same thing with the reality of the existence of God. Imagine someone being so arrogant as to say, "No, forget it. I'm not even going to look at what you're saying; you have to prove to me these things are real. That's ridiculous! Unless I can see them, feel them, taste them, touch them, and hear them in their pure essence, there's no way I'm going to believe it." That person's going to miss out on a whole lot. I mean, come on—imagine that. Missing out on CNN and Fox News, I mean, give me a break. How miserable life would be.

It's the same thing with the existence of God, folks. Listen, you've never seen Him or felt Him or touched Him or heard Him. You need the instrument that will reveal to you His reality, and that instrument is the Holy Spirit, and He is given to you by repentance and faith in the death and resurrection of Christ. So I pray you all will listen to what you've heard today, turn from your sin, and place your faith in Christ. Thank you so much for your attention.

* * *

Transforming Touch

I know that some of you standing here this morning have several thoughts regarding Christianity, and Christians specifically. All of you know that Christians claim to love God. That's our heartbeat. That's our passion in life. That's why we're here, because we love God and we want to obey Him, and because we love people. But I know that in some of your minds you think, *How in the world can you love Someone, and be so excited about Someone that you've never seen?* It says in the book of First Peter, regarding Jesus Christ, it says, "Whom having not seen you love. And though you do not see Him now, you rejoice with joy inexpressible and full of glory." That's so accurate in relation to the Christian. I've never seen God, and I don't see Him now, but I love Him with all my heart, and when I think about Him I rejoice with joy that's inexpressible and full of glory. There's a reason for that.

I liken it to someone that all of us are familiar with, a historical figure that is considered to be remarkable in the eyes of everyone. All of us have heard of Helen Keller. When you think of Helen Keller you think of a person who was, if you know her testimony, out of control. She was deaf and blind—couldn't see anything, couldn't hear anything. Her parents had tried time and again to get her help—someone who could kind of help at least get her under control, but no one could. Everyone gave up on her, so she was mad. They couldn't even seat her at the dinner table. She'd throw things across the table. She'd go wild and crazy.

But one day, someone entered her life who was willing to give it her all, to see her change. Her name was Anne Sullivan. Anne Sullivan came into Helen Keller's life and persevered with her until she finally broke through, and that day came when suddenly, in a sense, Helen Keller's eyes were opened, if you would, and she was transformed. She became the most remarkable . . . one of the most remarkable figures in history. I mean, she went on—imagine being deaf and blind, not able to see, not able to hear—she went on to become a lecturer, an author, an inspirational, motivational individual who touched people around the world.

Now, here's the thing that amazed me. I remember when I first realized this, it blew my mind. Think about this: Helen Keller never once saw the face of Anne Sullivan. She never once heard her voice, but I can guarantee you that there was no one on earth that Helen Keller loved more than Anne Sullivan. You can guarantee that there's no one on earth that Helen Keller was more excited about than Anne Sullivan—a person whom she had never seen and whose voice she never heard. And you know why? When you can't see or hear, the only other sense you can experience another individual through best, is the sense of touch. Helen Keller loved Anne Sullivan more than anyone on earth and was more excited about her than anyone on earth, because Anne Sullivan touched Helen Keller's life in such a way that transformed her, that she couldn't help but have such affection toward her.

And, friends, it's the same way for the Christian. Yes, we've never heard the voice of God. We've never seen His face, but He has so transformed our life. He has so revolutionized us that you can guarantee we so love Him and are so excited about Him. And that's why we're here this morning—to tell you that God is able to transform your life, that God is able to touch you in such a way that you could be revolutionized forever. But our hope is this morning that your eyes will be opened to realize that God is waiting to touch you, but you have to recognize your need for Him. You have to understand your desperate need to repent of your sin and recognize that the only hope that you have is the Son of God, whom God sent to hang on a cross for your sins and to rise again from the dead.

And you think about that. Oftentimes we think that God owes us something as people. He does—He owes us His wrath and judgment and justice, for all eternity in Hell. But think about that. Instead of that, He's giving us an opportunity to be delivered from that. But it's not like God one day just sat in Heaven and said, "Okay, you know what? I'm going to just give you the opportunity to be forgiven." Something had to happen, because the Bible says that without the shedding of blood there's no forgiveness of sin. And so, God had to become a man two thousand years ago, walk this earth, be ridiculed, mocked, tortured, and then murdered on a cruel cross, in order to give us life. And then He rose again on the third day from the grave so that we can be born again through His resurrection from the dead.

Think about that. Can you imagine today becoming a Chihuahua to save a breed of ravenous pit bulls and Doberman pinschers that you knew were going to devour you? And you did it anyway to save them. I mean, that's preposterous, but God condescended even more than that when He became a man to give us eternal life.

I remember watching *The Passion of the Christ*. I'm a pastor as well, and I've taught sermons on the crucifixion, and I've studied it historically. When I saw what I believe was this accurate depiction historically of the crucifixion again, I stood back and I asked myself, "God, why? Why did You do it? You didn't have to." And again the resounding answer came, "Because of My love for man." Who can reject such love? Who can spit on it as though it were nothing?

I want to urge you today to recognize that you're not just a good person who's kind of walking through life, but you're a sinner who's broken the Law of a holy, righteous God who's given you every gift that you enjoy. The sense of sight, and sound and smell and taste and touch; the ability to love; the blessing of your children, and your family, and your loved ones. God has given you all this, and yet you're separated from Him because of your sin, and today He's telling you that you can be reconciled to Him again. Join back together to Him again by simply humbling your heart, repenting of your sin, and believing in the death and resurrection of Christ, and declaring Him as Lord.

So I urge you to do that today, folks, and again, I thank you so much for your attention. I know you didn't even have to so much as look our way today. As you lay your head down on your pillow tonight, remember you've heard the truth and it's by no accident. What you do with it is up to you, and I urge you, please, do something about it today.

* * *

True Value

I know that some of you hear what we're sharing this morning, and in a sense it's worthless to you. You kind of tune out. You're hearing a little bit of what Charlie Brown used to hear on those episodes in the classroom, "*Wha wha, wha wha, wha wha.*" That's all that's kind of penetrating your mind. There's a reason for that.

I remember several years ago picking up a copy of *Time* magazine and flipping through the pages, when my attention was captured by two words. The first word was "Bill" and the second word was "Gates." I had enough sense to know that those two words added together, at least back then, equaled about $40 billion, and so it captured my attention. It was an article about the $100 million mansion that Mr. Gates was erecting on Lake Washington. This thing was incredible!

I'm a gadget freak, so I was really intrigued with all the details about the gadgetry he has around his house—along with the underground parking structure for his custom automobiles, and the private beach, and the indoor gym, and the maid's quarters. I mean this thing was spectacular. But there was one thing in particular that really captivated me. It was the fact that Mr. Gates was placing a specific item in the library of this palatial home that was worth $30.2 million. This item was the original sixteenth-century sketchbook of Leonardo Da Vinci. Get this—$30.2 million for an old, dusty, ragged, tattered sketchbook that has the scribbling of some guy who died several hundred years ago. You can't wear it, you can't fly it, you can't drive it, you can't even eat it! Thirty point two million dollars! Even for Mr. Gates, that's a lot of money!

And so the question popped into my mind, naturally, "Why in the world would anyone pay such a high price for something like that?" And the logical answer, of course, is because Mr. Gates understands the true value of the sketchbook. Its historical, its artistic, and its monetary value, and he understands that it's an investment, and in years to come it'll be worth double that, triple, quadruple that.

I had an interesting thought as I was considering this. I thought, *You know, if I brought this sketchbook home to my four children, who are eight years old and under, and at the same time I went to Big Lots and purchased a coloring book with a pack of crayons for about a dollar fifty, and I brought these home to my kids and I put them on the table, one next to the other and I said, "Hey, kids, guess what? You're going to get a gift from me today, but you have the choice of one or the other. What do you want?"* Do you think that they'd be interested in the old ragged, tattered, dusty sketchbook? Or do you think they'd be interested in that pack of crayons and coloring books? You can guarantee they'd discount that notebook. In fact, when they finish coloring their coloring book, they'd take that sketchbook and color it from page to page, and what's the reason for that? It's because they lack the understanding of the true value of that masterpiece. If they had the proper understanding, they would realize that this sketchbook holds their future. I mean, this will secure them. They'll never have to work a day in their life. Neither will their children, nor grandchildren, nor great-great-grandchildren. But because they lack the understanding, they treat it as though it were worthless. And they consider something that really is practically worthless of higher value in comparison to it. And, friends, the same thing happens in the spiritual sense.

What we're talking to you about today is the only means that God has provided for your soul to be saved from the eternal flames of Hell. And some of you today are treating what we're sharing with you like an old, worthless coloring book and pack of crayons, because you're not realizing the true value of what we're sharing with you. I want to urge you this morning to recognize that you've offended a holy God. That might not seem like a big deal to you, but it's a big deal to Him. The reason we know that it's a big deal is because of the price He paid to give us forgiveness of sin: by sending His Son to hang upon a cross, to shed His blood for us and to rise again from the dead that we can be forgiven. That's a high price to pay. God becoming a man, and enduring the mockery, and the ridicule and the blasphemy, and the torment, and torture of His very creation. I urge you today to wake up and realize the value of Jesus Christ, that even though His name is worthless to you now, it would become the most precious name in the world.

If I walked up and down this line this morning in light of the analogy I just shared, and I held two glasses in my hand—one glass was filled with diamonds, retrieved from the mines of South Africa, worth millions of dollars, and in the other hand I had a glass filled with mountain spring Evian water—and I said, "Hey, you have the choice of one or the other this morning—which would you like?" It'd be kind of childish to truly ask that question. But imagine if I were to sort of switch up the circumstances.

Let's say you were lost in the Sahara Desert, having not had a drink of water in days—literally on the brink of death—and suddenly I miraculously appear out of the blue, the first sight of human life you've seen in a week. And I hold in front of you that same glass of water and that same glass of diamonds and I tell you, "You have the choice of one or the other. You have five seconds to make your choice." Then I disappear. You can guarantee you'd despise those diamonds and grab that water with all you have, because you realize that it's your only source of survival.

So, folks, flee from the wrath to come, turn to Christ and let your soul be saved. Thank you so much for your attention again, we so appreciate it. If you'd like one of these CDs as we come along, please feel free to take it. It's called "What Hollywood Believes." It will tell you about the intimate beliefs of Hollywood celebrities. God bless you.

⋆ ⋆ ⋆

Who Can Tell Me?

I've often been amazed at the fact that, even though we as human beings are intelligent enough to recognize that certain things could have never come into existence on their own, we fail to recognize the same thing about ourselves. Especially when you think about the fact that we are the most complex organisms on the face of the earth. Do you ever stop for a moment to think about this incredible machine that houses your spirit? I mean, think about that. Your eye has 137 million light-sensitive cells. The focusing muscle in your eye moves an estimated 100,000 times a day. Your eye has a built-in super-sensitive light meter, immediate automatic focusing, wide-angle lens, and full-color instantaneous reproduction. Darwin himself said about the eye, "To suppose that the human eye could have been formed by natural selection seems I freely confess absurd in the highest degree." Think about that. Your brain has 10 to 20 billion neurons or microscopic nerve cells. It's got many times more nerve lines than all the telephone lines in the world put together. The electrical signals from 200,000 living thermometer cells, a half million pressure-sensing cells, and 3 to 4 million pain-sensing cells. Plus all the signals from your eyes, ears, nose, and taste buds are all routed to your brain that keeps every single thing in your body working in perfect order and harmony. Think about your heart. Your heart, like the focusing muscle in your eye, pumps 100,000 times a day, moving 75 gallons of blood through your body every hour with a work expectancy of 75 years, without ever shutting down for maintenance and repair. I mean, folks, that's incredible.

The thing that amazes me, though, is that instead of standing back and being in awe of God in light of these incredible features that He's given us, we use them instead to spite Him. Instead of thinking, "Wow, I'm able to see right now. I can see imagery and color, and my brain takes it and processes it. I understand it. It stores it. I remember it." Instead, we use our eyes to lust and look at filthy things, to despise the God who gave us life. Instead of thinking, "Man, what is the sense of sound? I'm able to hear words right now. My brain translates them. I comprehend them. I'm able to respond." Instead of doing that, we use our ears to listen to things that are displeasing in the sight of God, and then repeat those words that are hateful and spiteful toward others. Instead we should stand back and say, "Man, I've been created. I've been designed. I've been made by a holy God." And why have I been made? To know Him and to live for Him forever.

There's a poem I wrote years ago. I'm going to share it with you. It's about this whole thing of where we came from.

"Who can tell me where I came from?" The little boy would ask.
His question was a good one yet he faced a trying task.
Each man had different answers as he was soon to learn.
This brought about confusion and it caused a deep concern.

He first went to his schoolmates and they spoke with one another.
Then the smartest little toddler said, "You came from your mother."
Now this had satisfied him, yet only for a time.
For as he grew in years of age his thoughts began to climb.

He then looked all around him at all that he could see.
Then his mind began to wonder how it all had come to be.
He thought about the cosmos, the infinitude of space.
And every star and planet that exists in every place.

He thought about the rounded earth spinning in rotation.
And all four seasons that occur in yearly circulation.
He thought about the darkness and he thought about the light.
He thought about the sun and moon that helped the day and night.

He thought of all the creatures of the land and sea and skies.
Of all the different species and their variance in size.
And filled with curiosity this boy would daily strive
In hopeful expectation that his answer would arrive.

So he spoke with scientific men who claimed his question solved.
They told him of a real big bang and that all things evolved.
He then spoke with philosophers and many did insist
That there is no reality and nothing does exist.

He spoke with many people, various groups and sects
And heard the vast opinions of various intellects.
Now baffled by confusion, a very troubled view
Unable to discern what is error, what is true.

He almost gave up looking but he took a second look
And very unexpectedly he found a special book.
As he opened up the first page the mystery came undone.
His questions all were answered in Genesis chapter one.
With a nod of understanding he smiled so elated.
For now he surely knew, in the beginning God created.

Folks, I want to urge you today to recognize that God created you, but you've sinned against Him. You've disobeyed Him. You've broken His Commandments, and so because of that you've been separated from Him, and have incurred the judgment of Hell. But remember the good news we want to leave you with: that God sent His Son to die upon that cross and shed His blood for your sins and to rise again on the third day. If you acknowledge your sins today and repent, you'll be reconciled to your Creator. You'll be forgiven of your sin and you'll be granted everlasting life. So please consider that today. Thank you again so much for your attention. We know you didn't even have to so much as look our way today. We so appreciate it.

* * *

What Will It Profit?

I know that many of you this morning have several things going through your mind as we share with you. I know some of you are holding onto this world. It's all that you know, and it's all that you think there is. You're not willing to let it go for the sake of eternity. Now I know logically speaking, even you realize that that's unreasonable. But the thing is that so often we have a hard time grabbing onto that which we can't see. But on the other hand there's so much that we believe in that we can't see, it's amazing. I assure you that in the end you will regret the fact that you held on to things that are going to fade away.

There's an interesting story that surrounds the death of the former European emperor Charlemagne. Legend has it that before Charlemagne died he gathered together all of his servants, and he gave them specific instructions to be carried out after his death. He said, "After I die I want you to put me in my tomb, sitting upright on my throne. I want you gather all of my treasure around me, put my royal mantle around my shoulders, and my scepter in my hand. I want you to place upon my lap a specific book opened to specific section.

Eventually Charlemagne died and the Emperor of Philo was reigning in his place a couple of hundred years later. So he sent his servants to go and see if Charlemagne's requests had been carried out, and sure enough they came back and reported that they had. But only now two hundred years later, that body that used to sit majestic and erect was now a pile of bones. That crown and that royal scepter had fallen to the ground. That royal mantle and all the treasure were filled with dust, but upon his skeletal thighs was that book he requested—the Bible—and they reported that Charlemagne's boney index finger was pointing to the portion of the Bible that says, "What will it profit a man if he gains the whole world but loses his own soul?"[562]

What a fitting picture that is. Friends, that's the question I pose to you today: What would it profit you if you were to gain the entire world—all the fame and wealth and power and popularity—and in the end you die and spend an eternity in Hell? It will profit you nothing.

I want to urge you today to think about what you've heard: that God offers you forgiveness, that even though you've broken His Commandments, even though you're deserving of His wrath and judgment, in His love and in His mercy He became a man 2,000 years ago, hung upon a cross and shed His blood for your sins, and rose again from the dead. Now if you repent, if you acknowledge your spiritual bankruptcy and your sinful and wicked heart, and ask Him to forgive you and save your soul, He will transform you, give you everlasting life. I want to urge you to do that today, and to see the great grace that God will extend to you. Thank you again so much for your attention. We so appreciate it. If you would like one of these CDs, it's free. It's called "What Hollywood Believes." It talks about the intimate beliefs of Hollywood celebrities. You're going to love it. Thank you so much. God bless you.

---

562 Mark 8:36

STUART SCOTT ("SCOTTY")

Introduction

Hi. My name is Scotty. And what a blessing it has been listening to the gospel preached by Ray and EZ every morning at the courts. I was eager to record them, knowing that it would be an encouragement to all of you as well. Listening to the preaching had the effect on me of making me hungry and thirsty to want to do the same. These recordings hopefully will do that for you, as well as the experience of open-air. It also will provide you with tremendous resource to draw on and to listen to over and over again.

As you know, we were stopped from preaching the gospel at the courts, and so it provided an opportunity at the local DMV. Ray and I had approached the manager, who told us that they would call the authorities and have us removed if we would try to do open-air. Add to this that the lines could be as long as 120 people, stretched clear around the building. Well, that can be intimidating. And when I had decided to make an attempt at open-air, I was very nervous and scared and intimidated—so much so that you might think that it was my first time though I've preached open-air many times.

I originally made the recording as a precaution thinking I might need it later. And none of my fears came about or were realized. I wasn't removed. Nobody complained. And, when I listened to it, I was reminded of something. Quite awhile ago I was in an evangelism class and my trainer would preach the gospel, and we were to listen and follow the example. But I remember one time listening to him and thinking to myself, *Well, I could do that.* And so when I listened to this recording I imagine some of you thinking the same thing: *Well, I can do that.* You see, I'm not Ray or EZ. I'm just somebody who's willing and has a care for the lost. And so I'm hoping that this recording will encourage you to realize that you can do it too.

One last thing: in doing open-air, I've realized that there are a couple of different aspects. There is the fear, and so it's a challenge to overcome your fear. And then when you do there is a great victory. And that's a real aspect to doing open-air. There is also the side of obedience. God commands us to preach the gospel. And when you do there is a nearness to Christ that you will never experience otherwise. But that can't be the motive. Though those are real aspects of open-air, the underlying foundation needs to be a care for the lost.

When I look at people I am reminded of my father. Sometimes looking at the back of somebody, I see my son. I think of my grandchildren. It could be a wife or a husband. They are somebody's wife or husband, brother, sister, mother, father. And if it was somebody that you cared about, your love for them would overcome your fear. And that's what Jesus did for us. He set Himself aside. He came and He cared for us. And that's what we're to do. We're to set ourselves aside.

Put your confidence in God. Have faith in God and in His promises. Ignore the fears of man. And just do it anyway. I hope that the recording that follows will encourage you to share your faith, and realize that it's not your ability but your willingness. And God is with you and He'll strengthen you and uphold you with the right hand of His righteousness.

The Mirror of the Law

Okay, *it is October 13, 2005. I'm going to the DMV. I don't know how this'll work, but I'm going to try to open-air. I've been coming here for well over a month, passing out tracts. There's just a small crowd this morning. But it doesn't matter. God is on me about this thing.*

Good morning, folks. Could I have your attention for just a moment? I'm not with the DMV. I'm just a citizen who works over here in Bellflower, and I want to talk to you about something really important. Some of what I may have to say may offend some of you, but before you object, I want you to keep in mind that over a thousand soldiers—our young men and women—have given their lives recently in Iraq, in foreign soil to protect the freedoms of this country. One of which is the freedom of speech. So that's what I'm exercising here this morning. And the reason I'm doing it, though, is not to be offensive, but because I have something of such great importance that I would risk your disapproval or your objections. And it's this:

Folks, ten out of ten are going to die. We're all going to die, and the Bible says that you're going to face God in judgment. Many people think, "Well, I'll deal with it when I get there." But if you wait that long it's too late. I want to show you something that will be absolutely crystal clear in your mind of what to expect when you get there on Judgment Day.

The Bible says we need righteousness. That's perfection. You can't have one sin and go to Heaven. But what is sin? You remember the Ten Commandments? I want to show you just a couple of them. The Ninth Commandment says, "You shall not lie." Have you ever told a lie? Ever in your life, even one? Everybody here would have to admit to that. But do you know the Bible says that all liars will have their part in the lake of fire? That's Hell.

Have you ever stolen anything, even if it was small? It wouldn't matter if I reached in your pocket and pulled out a dollar or a hundred dollars; it would be theft. Have you ever stolen something, ever in your life, even once? That would make you a thief.

You know that it's wrong to commit adultery. That's to cheat on your wife or your husband, but Jesus has said whoever looks with lust has committed adultery already with that person in their heart and in their mind.

God even sees your thought life. And everything that you think you did in secret is open to God. And so, God says this: It's appointed for us to die once and then the judgment. You're going to stand before God, and that's how He's going to judge you. It's not my standard or yours. You may think that you're a good person compared to everybody else. But if God judges you by that measure, by that standard, if that's His determination of right and wrong, good and bad, then you'll realize that you're liars, thieves, adulterers at heart in the eyes of God.

The Law was given like a mirror so that you could see yourself in the true light of God's eyes—by *that* standard—so that you would realize you have no hope of entering Heaven on the day you die, if that's how He's going to judge you. But the good

news is that 2,000 years ago, God, though we were enemies to God in our minds by wicked works in the things that we do against God, He came down in the form of a man. He was born of a woman, and He never lied, never stole; He never had a lustful thought—this man Jesus. When He was accused, He was absolutely innocent, but He willingly went to the cross and bore the shame that was our shame. And when He hung on the cross, the Bible says that God took all of the sins of the world, all of our lies and thefts and adulteries and put them on His Son, Jesus Christ, and that He died for our crimes. We broke the Law, but Jesus paid the fine. This is the mercy of God, and this is where forgiveness is found.

Greater love has no man than this, that a man would give his life for his friends. And God says, "If you will return to Me, if you will repent and turn back to God and put your faith in My Son, Jesus Christ, then I will wash your sins away. I will give you the gift of eternal life."

And so all that's required of us who are condemned in trespasses and sins and on our way to Hell and have nothing we could do about it, is to return to God. So repentance means to turn 180 degrees—turn toward God. Believe Him and believe in His Son, Jesus Christ, just like you would a parachute. You wouldn't jump out of an airplane with the parachute sitting by the door and not put it on, would you? I mean, that's ridiculous. But that's what we're doing when you're going to face death and Judgment Day without the Lord Jesus Christ. Put Him on like a parachute. That means to follow Him.

Folks, you probably all have a Bible, and if you don't have one you can get one. Read it, obey what you read. God will never let you down and you'll never be ashamed for putting your trust in Him. He is faithful and true and He has done all that He can to save you from your sins. But if you reject it, you'll die in your sins and you'll answer to that standard. Please don't do that. Please think about these things today while you're alive, because when you die your choice is over.

Thank you so much for listening to me. You see why I felt it important enough to say something to warn you. Escape the wrath to come and put your faith in Jesus Christ. I have some literature I just want to give you—anybody who wants it in Spanish or English. And thanks again for listening.

Thank You, Jesus. Thank You, Lord…Ahh…that was intimate. It was like maybe twenty people, and they all listened. And they all took tracts except for one person. Oh, Lord, help them. Open their eyes in Jesus name, Amen.

You know, I've been so worried about doing this…anxious, and we all know that's not of God. It's a lack of trust, but all this week for a whole week Isaiah 58:1 has been sitting in front of me as I'm reading through Isaiah and just kind of the way things worked out…just stayed camped there for a while, and it said something like, "'Cry out,' says the Lord. 'Lift up your voice like a trumpet and convict My people of their transgression and Jacob of their sin.'" Lift up your voice like a trumpet…Cry aloud…Convict them. That's exactly what we do in open-air. And so I thank You, Lord, that You brought this about. It was small and it was easy. I didn't have to yell or stand across in the parking lot or speak to a crowd of 80. And God, You are always so faithful and make a way when we're timid and frightened and scared, and all of our fears are unfounded as we follow after You. Thank You, Jesus.